Richard Burleigh Kimball

Student's Abroad

Richard Burleigh Kimball
Student's Abroad
ISBN/EAN: 9783337138684
Printed in Europe, USA, Canada, Australia, Japan
Cover: Foto ©berggeist007 / pixelio.de

More available books at **www.hansebooks.com**

Romance of Student Life Abroad

ABROAD.

BY
RICHARD B. KIMBALL,
AUTHOR OF "ST. LEGER," "UNDERCURRENTS" ETC.

NEW YORK:
G. P. PUTNAM, 532 BROADWAY.
1862.

Entered according to Act of Congress, in the year 1853,
BY RICHARD B. KIMBALL,
In the District Court of the United States for the Southern District of New York.

The Ancient Art rigorously separates things which are dissimilar; the ROMANTIC delights in indissoluble mixtures, all contrarieties: nature and art, poetry and prose, seriousness and mirth, recollection and anticipation, spirituality and sensuality, terrestrial and celestial, life and death, are by it blended together in the most intimate combination.

The Ancient Art is an harmonious promulgation of the permanently established legislation of a world submitted to a beautiful order. The ROMANTIC is the expression of the secret attraction to Chaos which lies concealed in the very bosom of the ordered Universe, and is perpetually striving after new and marvellous births.

The former is more simple, clear, and like to nature in the self-existent perfection of her separate works; the latter, notwithstanding its fragmentary appearance, approaches more to the secret of the Universe. For Conception can only comprise each object separately, but nothing in truth can ever exist separately and by itself; Feeling perceives all in all at one and the same time.—A. W. VON SCHLEGEL.

CONTENTS.

CHAPTER I.

A first adventure.—Calais.—A new acquaintance.—Mr. Philip Belcher.—His theory of travel.—He proffers good advice.—He gives an account of himself.—Story of Louis Herbois.—New visions.—New prospects.—New anticipations. . . 17

CHAPTER II.

A surprise.—A discussion.—The diligence.—Capitaine Duclos.—Entertaining women.—Beautiful hamlets.—Fantastic graveyards.—Images of the Virgin.—Signs over the different Hotels.—Donkeys.—Dinner *en route*.—Partridge recites poetry.—Scene at the Inn.—The bougies.—Partridge turns sympathizer.—Hotel Sauvage.—Hotel des Gentilhommes.—We

enter Paris.—We look about.—Visit to the Student's Quarter.
—Monsieur Battz.—The Mademoiselle Battz.—Our new set.—
"Walking" the hospital.—The young English doctor.—His
peculiarities.—Man and woman discussed. 56

CHAPTER III.

Clements illustrates.—Students in Paris.—Students in Germany.
—The distinction.—Habits of the former.—The Story of
Ludwig Bernhardi. 76

CHAPTER IV.

Rambles over Paris.—Charbon and fagot venders.—Jacques
Tourneau.—The gardens.—Hotel des Invalides.—Old soldier
with two wooden legs.—The chapel.—Old soldiers at prayers.
—The melancholy officer.—Light and shadow.—Incident in
the chapel.—Children playing.—Little Annie.—Her grandmother.—French delicacy.—An affecting scene. . . . 102

CHAPTER V.

Students' nonsense.—After dinner.—Our company.—Daloney.—
Franz Vor. Herberg.—Jacob Wahlen.—The two Englishmen.
—Vincent.—A good shot.—The picture Franz cannot paint.—
Putting two things together.—The new hat.—The juggler.—A
dangerous suggestion.—National characteristics.—We visit

an artist's room.—Its appearance.—The wrong painting.
—The whole party struck with horror.—We beg for an explanation. 111

CHAPTER VI.

Life not a particular form of body, but body a particular form of life.—Story of the Terrible Picture.—Vincent feels unsettled.—His invitation.—Champagne.—Pipes.—Meerschaums and segars in requisition. 121

CHAPTER VII.

Vincent proposes to tell a story.—He makes an inquiry in advance.—It is answered.—He requests the company not to be impertinent.—He tells the story of the Water-carrier.—The company break up in fine spirits.—Nobody thinks of the terrible picture 136

CHAPTER VIII.

Mornings at la Morgue.—Melancholy sights.—The pale woman.—Young girls.—Young men.—The little child in search of "Mamma."—The old man.—Leave Paris.—Return in the summer.—Jardin des Plants.—Partridge proposes a remarkable enterprise.—We attend at the rendezvous.—The strange ap-

pearance.—An hour of suspense.—Partridge explains.—Story
of the Fair Mystery. 173

CHAPTER IX.

Changes.—The rue Copeau abandoned.—A haunted house.—The
Italian.—He refuses to enlighten us.—Vincent reads a letter.
—A melancholy Jacques.—An account of New York society.
—The Italian discourses about physicians.—He will go to
America.—He tells a strange story of a dead man on the
Boulevard.—The dispersion. 205

CHAPTER X.

New quarters.—Franz Von Herberg.—Rue de la Chaussée d'Autin.
—Our opposite neighbours.—The backgammon players.—The
two grisettes.—Mother and idiot son.—Shopkeeper's family.
—People of fashion.—French economy.—What Franz tried
to paint.—Our Lady of Lorette.—Old mendicant.—His death.
—A serious discussion.—Franz is in doubt.—Champaux's. . 216

CHAPTER XI.

The café.—A character.—The garçon is puzzled.—He wears a
permanent shrug.—He is in despair.—We proceed to his
assistance and discover an acquaintance.—Wilcox gives an
account of his efforts to keep from starving.—New method

of dining.—Wilcox on his travels.—He reaches Lyons.—He attempts to go to Marseilles.—He gets into trouble, and then into prison.—Meets with fresh misfortunes.—Is at last set at liberty.—A funeral scene.—Death.—Mourners.—The artificial and the natural. 224

CHAPTER XII.

Almost at the end.—We tire of fashionable quarters.—Partridge returns.—We prepare to leave Paris.—Franz's new painting. —Partridge is inquisitive.—The story is demanded.—It is insisted on.—It is told.—Story of Marie Laforet. . . . 238

CHAPTER XIII.

Preface for conclusion.—Author and literary friend.—A mistake which cannot be corrected.—Publisher shakes his head.—A compromise. 258

ROMANCE OF STUDENT LIFE ABROAD.

CHAPTER I.

A FIRST ADVENTURE.

We intended—Partridge and myself—to go directly from Liverpool to Paris. It is what most youth decide to do when they find themselves for the first time on European soil. But we reconsidered the matter. After enjoying with a keen relish the comforts of an English inn for twenty-four hours, we concluded to make a tour of Great Britain and Ireland, before settling down to study.

This was several years ago. It is hardly prudent to count back how many. On second thoughts, I resolve to do it: I now say, with accuracy, it was sixteen years.

At that period a voyage across the Atlantic was a thing to be remembered for a life-time. Lasting friendships were formed, and often what were more significant than friendships; for many were the vows to which Neptune was the witness, and frequent their interchange, on the decks of our magnificent packet-ships, those fine nights,

> "———— While overhead the moon
> Sits arbitress, and nearer to the earth
> Wheels her pale course————"

In short, it was like a charming visit of a month at the mansion of some hospitable friend, whose abode is filled with a large and congenial company. How all this is changed! The idea of wooing and winning a lovely maiden on board of a steamer, while the engines, impelled by

> "Tartarean sulphur and strange fire,"

beat time to your protestations with their clack—clatter—clank! Alas, the friendly mansion is converted into a noisy hotel, and the visit of a month reduced to a stay of ten days.

On the other side of the ocean, the mail-coach and diligence were in their glory, while at almost every turn of the road one encountered the post-chaise or calêche of the private traveller. I mention this by way of parenthesis, and proceed to remark that, after making our proposed excursion through England, Ireland, and Scotland, and as we were about to cross the Channel from Dover, my companion missed his portmanteau and was obliged to go back to London in search of it; while I, eager to get into France, passed over to Calais, promising to wait for him there.

I shall give no account of my friend's exploits in pursuit of his lost luggage—I shall not even tell whether he found it or not. I am to speak of this, my first adventure into France, and how I fared at Calais. The landing and getting through the custom-house, the examination of passports and so forth, diverted me for a few hours. The aspect of every thing around—to me new and peculiar—made the following day pass cheerfully enough. On the third, I attempted a drive on the road to St. Omer, and returned covered with dust, without seeing a single object to interest me. It was now with difficulty that I could occupy the time. As a last resource, I took to inspecting the different faces which daily presented themselves at the Hotel de Meurice, where one could see a great variety of features, belonging to almost every country, age, sex, and condition. But I tired of this presently, so that when the fifth day brought with it one of those disagreeable storms peculiar to the coast—half drizzle, half sleet and rain—it found me weary of the amusement of attending on new arrivals and departures, and of the nameless petty doings by which time, in a bustling hotel, is attempted to be frittered away. A misty, dreary, damp, offensive day! An out-and-out tempest, a thorough right down drenching rain, would have been in agreeable contrast with the previous hot, dusty, sunny weather; but this—it seemed absolutely intolerable! I was, besides, in no particular condition to be pleased. I was neither setting out upon a tour, nor returning from one, but had been interrupted in my progress and forced, with loss of my companion, to a stand-still at this most uninteresting spot. I came down, and with a bad grace, to order breakfast.

"Garçon, Café—œufs a la coque—biftek—rotie—vite!"

I was about repeating this in a louder tone, for the waiter seemed engrossed with something more important than attending to my wants, when I heard a quiet voice behind me—

"Garçon, Café—œufs a la coque—biftek—rotie—vite!"

I turned angrily upon the speaker, doubtful of the design of this repetition of my order.

The reader will perceive that my breakfast was a substantial one; indeed, such a breakfast as an American, who had not so far lost himself in "European society" as to forget his appetite, would be very likely to call for. The idea that I was watched, doubtless made me a little suspicious, or sensitive, or irritable; at any rate, I turned, as I have said, angrily upon the speaker. He was a slightly made, elderly man, at least fifty, with pleasant features, a calm appearance, and quiet manners—a person evidently at home with the world. I recollected at the same moment, that the stranger had been at the hotel ever since my arrival there, although I had not, from his unobtrusive habit, given him more than a passing notice. His appearance at once dispelled the frown which I had brought to bear upon him;

but when he answered my stare with a respectful yet half familiar bow, I could have sworn that it came from an old acquaintance. I need not say that I returned the salutation cordially. At the same time my new friend rose, came towards me, and held out his hand.

"I am quite sure," he said, " that you are an American —perhaps a New Englander; I am both; why, then, should not countrymen beguile an unpleasant day in company? Excuse me—I did hear your order just now, and as it suited my own taste, I proposed to myself that we should breakfast together;—we may trust to François; he has been here, to my knowledge, more than twenty years, and pleases every body."

I pressed the hand of my new acquaintance—acknowledged myself to be from New Hampshire—gave my name, and received in return—" Philip Belcher."

We sat down to the same table, and very soon François appeared with a well-served breakfast.

" Pray," said I, " what *can* one do to relieve the monotony of this intolerable place? If the country about were agreeable—nay, if it were bearable! but as it is, I repeat, what is to be done?"

" Done !" said Mr. Belcher, rather sharply, " a hundred things! Put on your Mackintosh and overshoes; come with me to the Courtgain, and see the fishermen putting to sea, their boats towed out by their wives and daughters; a sight, I will be bound, you have not beheld, although you may have coursed Europe over, and been at Calais half a dozen times."

Mr. Belcher proceeded in this vein, detailing many things that could be seen to advantage even in Calais; but as he suggested nothing which interested me so much as he himself did, I had the boldness to tell him so, and that my curiosity was excited to know more of him.

"There is nothing in my history that can amuse a stranger; indeed, it is without incident or marvel. To be sure, I am alone in the world, but I have never been afflicted or suffered misfortune, within my recollection. My parents died when I was very young; my father and mother were both only children; a small property which the former left was carefully invested, and faithfully nursed during my minority, by a scrupulous and honest attorney, in no way connected with us, but whom my father named as executor in his will, and my guardian. Ill health prevented my getting on at school. I cannot say that I was an invalid, but my constitution was delicate and my temperament nervous. I tried to make some progress in the study of a profession, under my excellent guardian, but was forced to give it up as too trying to my nerves. The excitement of a court-room I could not endure for a day, much less for a lifetime. Before I was twenty-five, my income had so much increased that I could afford to travel. I have gained in this way my health, which, however, would become impaired should I return to a sedentary life; so, as a matter of necessity, I have wandered about the world. You see, my story is soon told."

I found Mr. Belcher was not in the habit of talking about himself, and I liked him the better for it. Without pressing for a more particular account, I led the conversation to treat of the different countries he had visited, referring, by the way, to some principal objects of attraction. Here I touched an idiosyncrasy of my new acquaintance.

"I never formed," he said, "any distinct 'plan' of travel. I never 'did' Paris in eight days, nor the gallery of the Louvre in half an hour, as they have been done by an acquaintance. I never opened a guide-book in my life; I never employed a *commissionere*, a *valet*, a *courier*, a *cicerone*, or a *dragoman*. My pleasure has been to let the remarkable, the beautiful, the interesting, burst upon me without introduction, and I have found my account in it. I have quitted the Val d'Arno, turned off from the Lake of Como, passed to the other side of Lake Leman and its romantic castles, pursuing my way, regardless of these well-worn attractions, while I beheld rarer—at least less familiar scenes, and enjoyed with zest what was fresh and unhackneyed. No everlasting 'route'—no mercenary and dishonest landlords—no troops of travellers, travelling that they may become 'travelled'—but, in place of all this, I saw every thing naturally—the country in its simplicity—the inhabitants in their simplicity—while, I trust, I have preserved my own simplicity. Indeed, I rather prefer what your tourist calls an 'uninteresting region.'"

"For that reason," I remarked, pleasantly, "you have come here to Calais to spend a few weeks; you must enjoy the barren sand-plain which extends all the way from this to St. Omer. How picturesque are those pollards scattered along the road, with here and there a superannuated old windmill, looking like an ogre with three arms and no legs! then, to relieve the dreariness of the place, you have multitudes of miserable cabins, grouped into more miserable villages, to say nothing of the chateaux of dingy red, in which painters of the brick-dust school so much delight. Really, Mr. Belcher, you will have a capital field here!"

My new acquaintance shook his head a little seriously, as if deprecating further pleasantry.

"You are like the rest of them, I fear," he remarked, "a surface traveller; at least you will force me to believe so if you go on in this way. To me there is no place unworthy of observation—no spot which does not challenge my attention. You are young, and have much before you. Take the advice of an old and, I trust, not an ill-natured traveller. Preserve the romantic in your heart, and you will never miss it by the wayside, no matter what you encounter, or how dull and disagreeable it may seem to another. But come," he continued, "I will not scold you; the storm threatens to last the morning; if you wish, I will help to make away with part of it, by recounting a little adventure which happened to me hard by those very pollards which you are pleased to abuse so freely."

It is needless to add that I joyfully assented to the proposal, and was soon seated in Mr. Belcher's room before a cheerful fire—for he had managed even in Calais to procure one—when he commenced as follows:

"I think it was during the first season I was on the Continent that I visited St. Omer. After spending a day or two in that place, I concluded to walk to Calais, and set out one morning accordingly.

"The weather was fine; but after I had been a few hours on the road, the wind began to blow directly in my face, and soon enveloped me in a cloud of sand from which there seemed no escape, and which threatened actually to suffocate me. To avoid this I left the highway, but keeping what I supposed to be in the general direction of the road, I struck out into the adjacent fields. There was nothing for a considerable distance to repay me for this *detour*, except that I was thus rid of the sand. The country was barren and uninviting, the cottages little better than hovels, and the whole scene distasteful. But I pushed on, not a whit discouraged; indeed, my spirits rose as the prospect darkened, and like a valiant general invading a country for the purpose of conquering a peace, I resolved in some way to force an adventure before I reached Calais. I trudged along for hours, stopping occasionally for a draught of sour wine and a bit of bread. I made no inquiry about the main road, for I preferred to know nothing of it. In this way I proceeded, until it was almost night, when I spied, some half a mile distant, a cluster of trees surrounding a small tenement. I turned at once toward the spot, and coming up to it, found a cottage not differing in size or structure from those I had seen on the way, except that it appeared even more antiquated. It was, however, in perfect repair, and finely shaded by a variety of handsome trees, and flanked on one side by a neat garden. The door stood open and I entered. There was no one in the room. I called, but received no answer. I strayed out into the garden and walked through it. At the lower end was a small enclosure covered over at the top as if to protect it from the weather, and fenced on each side with open wire-work, looking through which, I beheld a small grave, overspread with mosses, and strewed with fresh-gathered white flowers. It bore no name or inscription, except the following simple but pathetic line·

'Enfant cherie, avec toi mes beaux jours sont passés.—1791'

Surprised by the appearance of fresh flowers upon a tomb which had been so long closed over its occupant, I turred, hoping to find some explanation of the mystery in what I might see elsewhere. But there was nothing near to attract one's attention, nor was any person within sight.

"After taking a glance around, I went back to the cottage, and walking in, sat down to wait the arrival of the occupants. In a few minutes, I heard voices from the side of the house opposite the garden, and soon two persons, of the peasant class, evidently husband and wife, came in. The man was strong and robust, with the erect form and martial appearance acquired only by military service, and which the weight of nearly sixty years had not seemed to impair. His countenance was frank and manly, and his step firm. The woman appeared a few years younger, while the air of happy contentment which beamed in her face, put the ordinary encroachments of time at defiance. Altogether, I had never seen a couple so fitted to attract observation and interest. They both stopped short on seeing me.

"I hastened to explain my situation, as that of a belated traveller, attracted by the sight of the cottage; and told them I was both hungry and tired, and desirous of the hospitality of their roof. I was made welcome at once.

"Louis Herbois, for that was his name, gave me a bluff, soldierly greeting, while Agathe, his wife, smiled her acquiescence. Supper was soon laid; I ate with a sharpened appetite, which evidently charmed my host, who encouraged me at intervals, as I began to flag.

"Supper concluded, I was glad to accept the offer of a bed, for I was exhausted with fatigue.

"I had been so engrossed with the repast, that curiosity was for the time suspended, and it was not again in action until I had said good-night to my entertainers, and found myself in the room where I was to sleep. This was an apartment of moderate size; the furniture was old and common, but neither dilapidated nor out of order; the bed was neatly covered; around the room were scattered several books of interest, and in one corner was a neat writing-desk, of antiquated appearance, with silver mounting, and handsomely inlaid; while some small articles of considerable value placed on a table in another corner, indicated at least occasional denizens very different from the peasant and his wife. Yet this could not be a rural resort for any family belonging to the town. There were but two other apartments in the house, and these were occupied. Nevertheless, I reasoned, these things can never have been brought here by the worthy people I have encountered; and then—the little grave in the garden? who has watched the tomb for so many years, preserving the moss so green and the flowers so fresh—cherishing an affection which has triumphed over time? How intense, how sacred, how strange must be such devotion! I decided that some persons besides those I had seen were concerned, in some way, in the history of the little dwelling, and with this conclusion I retired; and so, being fatigued by my day's travel, I soon fell asleep.

"I awoke about sunrise. Going to the window, I put aside the curtain, and looked out into the garden. Louis Herbois and his wife were there, renewing the garlands with fresh flowers, and watering the moss which was spread over the grave. It must be their own child, thought I, and yet—no—I will step out and ask them, and put an end to the mystery. I met the good people coming in: they inquired if I had rested well, and said that breakfast would soon be ready. 'You do not forget your little one,' I said to the old fellow, at the same time pointing towards the enclosure. 'Monsieur mistakes,' replied he, crossing himself devoutly. 'Some dear friend, I suppose?' He looked at me earnestly: '*On voit bien, Monsieur, que vous ôtes un homme comme il faut.* After you have breakfasted, you shall hear the story. 'Ah, there is, then, a story,' said I to myself, as I followed Louis Herbois into the cottage, where Agathe had preceded us, and sat down to an excellent breakfast. When it was concluded I asked for the promised narration. 'Let me see,' said Louis, 'Agathe, how long have we been married?' Agathe, matron as she was, actually blushed at the question, yet answered readily, without stopping to compute the time. 'Yes—true—very well;' resumed Louis. 'You must know, Monsieur, that my father was a soldier, and enrolled me, at an early age, in the same company with himself. Having been detailed, soon after, on service to one of the provinces, I was so severely wounded that I was thought to be permanently unfitted for duty, and was honourably dismissed with a life pension. Owing to the care and skill of a famous surgeon who attended me, and whom I was fortunate enough to interest, I was at last cured of my wounds, and very soon after I wandered away here, for no better reason, I believe, than that Agathe was in the neighbourhood; for we had known each other from the time we were children. Very soon she and I were married, and we took this little place, and were as blessed as possible.

"'In the mean time, great changes were going on at Paris. The revolution had begun, and soon swept every thing before it. But it did not matter with us. We rose with the birds, and went to rest with the sun, and no two could have been happier: am I not right, Agathe?' The old lady put her hand affectionately upon the shoulder of her husband, but said nothing. 'And we have never ceased being happy, we are always happy; are we not, Agathe?' The tears stood in Agathe's eyes, and Louis Herbois went on. 'Well, the revolution was nothing to me; they were mad with it, and killed the king, and slew each other, until our dear Paris became a bedlam—still, as

I said, it was nothing to me. To be sure, I went occasionally to Calais, where I heard a new language in every body's mouth, and much talk of *Les hommes suspects, Mandats d'arrets*, with shouts of *A bas les aristocrates*, and *Vive la Republique*—but I did not trouble myself about any of it; Agathe and I worked together in the field, and in the garden, and in the house—always together—always happy. One morning we went out to prune our vines; the door of the house was open, just as you found it yesterday; why should we ever shut the door? we were honest, and feared nobody; we stood—Agathe here on this side holding the vine; I, with my knife, on the other side, bending over to lop a sprout from it; when down came two young people —lad and lass—upon us, as fast as they could run, out of breath—agitated—and as frightened as two wood-pigeons. The young man flew to me, and, catching hold of my arm, begged me, *pour l'amour de Dieu*, to secrete his wife somewhere—anywhere—out of the reach of the *gens-d'armes* who were pursuing them. I felt in ill-humour, for I had cut my finger just then; besides, I did not relish the mention of the *gens-d'armes :* so I replied plainly, that I would have nothing to do with persons who were *suspects*. Why should I thrust my own neck into the trap? they had better go about their business, and not trouble poor people. Bah! such a speech was not like Louis Herbois! but out it came, Heaven knows how, and no sooner had I finished than up runs the young creature, and, seizing my moustache, she cries, "My brave fellow, hie away, and crop off all this; none but *men* have a right to it; God grant you were not born in France; no Frenchman could give such an answer to a man imploring protection for his wife. Look at my husband—did he ask aid for himself? Do you think he would turn you off in this way, had you sought his assistance to save *her?*" pointing to Agathe, who stood trembling all the while like an aspen. "Ah! you have made a mistake—I see you repent—be quick; what will you do with us?" And she held me tight by the moustache until I should answer, while the husband stared upon me in a sort of breathless agony. I took another look at the little creature, while she kept fast hold of me, and saw that she was——*eh bien!* I see you understand me,' said Louis, interrupting himself, as he glanced towards his wife. 'My heart knocked loud enough, believe me, and there the little thing stood, her hand, as I was telling you, clenched fast in my moustache—ha! ha! ha! —and looking so full into my eyes, with her own clear bright blue gazers. "*Mon Dieu—mon Dieu!* Agathe, we must help these *pauvres enfans.*" "You *are* a Frenchman —I thought so," cried the little one, letting go my moustache and clapping her hands. "Oh! hasten, hasten, or we are lost!" "All in good time," said I, "for—" "No, no," interrupted she, "they are almost upon us: in a moment we may be captured, and then, Albert, oh! Albert, what will become of you?" So saying, she threw her arms about her husband, and clung to him as if nothing should part them. "*Voila bien les femmes;* to the devil with my caution; come with me, and I will put you in a place where the whole Directory shall not find you, unless they pull my cottage down stone by stone." I hurried them to the house and hid them in a private closet which, following out my soldier-like propensities, I had constructed in one end of the room, in a marvellously curious way. Not a soul but Agathe knew of it, and I disliked to give up the secret; but I hurried the young people in, and arranged the place, and went back to the vines and cut away harder than ever. In two minutes, up rode three dragoons with drawn swords, as fine-looking troopers as one would ask for. I saw them reconnoitre the cottage, then, spying me, they came towards us at a gallop. "What have you done with the Comte and Comtesse de Choissy?" said the leading horseman. "You had better hold your tongue," I retorted, "than be clattering away at random. What the devil do I know of the Comte and Comtesse de Choissy, as you call them?" "Look you," said the dragoon, laying his hand on my shoulder, "the persons I seek are escaped prisoners; they were seen to come in the direction of this cottage; our captain watched them with his glass, and he swears they are here." "And look you, Monsieur Cavalier, I am an old soldier, as you see, if scars and hard service can prove one, and it seems to me you should take an old soldier's word. I have said all I have to say; there is my house, the doors are open—look for yourself: come, Agathe, we must finish our morning's work." So saying, I set at the vines again. I looked neither one way nor the other, but kept clipping, clipping, thus standing between the dragoons and poor Agathe, who was frightened terribly, although she tried to seem as busy as I. The rider, who was spokesman, stared for a minute without saying a word, and then broke out into a loud laugh. "An old soldier indeed!—a regular piece of steel!—one has but to point a flint at him, and the sparks fly." He turned to his men: "Our captain was mistaken, evidently; this is a *bon camarade;* we may trust to him. We will take a turn through the cottage and push forward." With that he bid me good-morning, and, after looking around the house, the party made off.

'" Well, Agathe, what's to be done now?" said I, when the dragoons were fairly out of sight. "We have made a fine business of it." "Ah, Louis," said she, "let us not think of the danger; we have saved two innocent lives, for innocent I know they are: what if we *have* perilled our own? Heaven will reward us." Nothing more was said, though we both thought a great deal, but we kept at our work as if nothing had happened. It was a long time before I dared let the fugitives come from their hiding-place; for I was afraid of that cursed glass of *Monsieur le Capitaine*. When I did open it I found my prisoners nearly dead with suspense. We held a council as to the best means for their concealment—for who would have had the heart to turn the young people adrift?—and it was finally settled that the Comte and his wife should dress as peasants, and take what other means were necessary to alter their appearance, that they might pass as such without suspicion. This was no sooner resolved than carried out. Agathe was as busy as a bee, and in a few minutes had a dress ready for Victorine—we were to call her by her first name—who was now as lively as a creature could be, running about the room looking into the glass, and making fun of her husband, who had in the mean time pulled on some of my clothes. After this, the young comte explained to me that his father had died a short time before, leaving him his title and immense estates, which, however, should he die childless, would pass to an uncle, a man unscrupulous and of bad reputation. This uncle was among the most conspicuous of the revolutionists. Through his agency the Comte de Choissy and his young wife, with whom he had been but a twelvemonth united, were arrested, and shortly after sentenced to death. They escaped from prison and the guillotine by the aid of a faithful domestic, and were almost at Calais when they discovered that they were pursued.

By leaving the road and sending the carriage forward, they managed to gain the few moments which saved them. Their principal fear now was from the wicked designs of the uncle, for the Directory had too much on their hands to hunt out escaped prisoners who were not specially obnoxious. For some days the young people did not stir from the house, but were ever ready to resort to their hiding-place on the first alarm. There were, however, no signs of the *gens-d'armes* in the neighbourhood. I went to Calais in a little while, and found, after much trouble, the old servant who was in the carriage when the Comte and his wife deserted it. He had been permitted to pass on without being molested, so alert were the soldiers in pursuit of the fugitives; and he had brought the few effects which he could get together for his master on leaving Paris to a safe place; and, to prevent suspicion, he himself had taken service with a respectable *traiteur*. By degrees, I managed to bring off every thing belonging to my guests, and we fitted up the little room, in which you passed the night, as comfortably as possible, without having it excite remark from any one casually entering it. "Albert" was industrious, aiding me at my work, no matter what I was doing, and "Victorine," too, insisted upon helping my wife in whatever she did, here, there, and every where, the liveliest, the merriest, the most innocent creature I ever set eyes upon. But for all that, one could see that time hung heavy on the

Comte. He became thoughtful and *triste*, and, like every man out of his proper place, he was restless and uneasy. Not so the dear wife: she declared she had never been so happy, that she had her Albert all to herself: she wanted nothing more: if she but knew how to requite *us*, she would not wish the estates back again—she would live where she was for ever. Then her husband would throw his arms around her, and call her by endearing names, which would make the little thing look so serious, but at the same time so calm and satisfied and angel-like, that it seemed as if the divine soul of the Holy Virgin had taken possession of her, as she turned her eyes up to her husband and met his looking lovingly down. . . .'

"Here Louis Herbois stopped, and felt for his handkerchief, and blew his nose until the walls resounded, and wiped his eyes as if trying to remove something that was in them, and proceeded:

"'Any one to have seen her at different times would have sworn I had two little women for guests instead of one: so full of fun and mischief and all sorts of pranks; so lively, running hither and yon, teasing me, amusing Agathe, rallying her husband; but on the occasions I mention, so subdued, so thoughtful—so different from her other self: *Ciel!* she had all our hearts.

"'Several months passed, much in the same manner. The Comte by degrees gained courage, and often ventured away from the house. Twice he had been to the town, but his wife was in such terror during his absence that he promised her he would not venture again. He continued, meanwhile, moody and ill at ease: it would be madness to leave his place of concealment; this he knew well enough; still he could not bring himself to be patient. Do not think, Monsieur, that the Comte de Choissy failed to love his wife just as ever: that was not it at all. A man is a man the world about; the Comte felt as any one would feel who finds himself rusting away like an old musket, which has been tossed aside into some miserable cockloft. I had seen the world, and knew how it was with him. But what could be done? In Paris things were getting worse and worse. At first we had *le Côté Gauche; les Montagnards; les Jacobines:* then came *les Patriotes de '93*; and after that, *les Patriotes par excellence*, who were succeeded by *les Patriotes plus patriotes que les patriotes:* and then the devil was let loose in mad earnest; for what with *les Bonnets-Rouges, les Enragés, les Terroristes, les Buveurs de Sang* and *les Chevaliers du Poignard*, Paris was converted into a more fitting abode for Satan than his old-fashioned country residence down below. *Pardon, Monsieur!* I am getting warm; but it always stirs my blood when I recall those days. I see, too, I am getting from my story. Well, I tried to comfort the Comte with such scraps of philosophy as I had picked up in my campaigns—for in the army, you must know, one learns many a good maxim—but I did little by that. The sweet young Comtesse was the only one who could make him cheerful, and smile, and laugh, and seem happy in a natural way, for he loved her as tenderly as a man ever loved; besides, the Comtesse had now a stronger claim than ever upon her husband. I fancy I can see her sitting *there*, her face bent over, employing her needle upon certain diminutive articles, whose use it is very easy to understand. Do you know, when she was at work on *these*, that she was serious—never playful—*always* serious; wearing the same expression as when she received from her husband a tender word! No; nothing could make her merry then. I used to sit and wonder how the self-same person could become so changed all in one minute. How the Comte loved to look at her! his eyes were upon her wherever she was; not a word she spoke, not a step she took, not a motion of hers escaped him. Well, the time came at last, and, by the blessing of God and the Holy Virgin, as beautiful a child as the world ever welcomed was placed by my Agathe in the arms of the Comtesse. Perhaps,' added Louis Herbois, in a lower voice, while speech seemed for the instant difficult, 'perhaps I have remembered this the better, because God willed it that we ourselves should be childless. When Agathe took the infant and laid it in its mother's bosom, the latter regarded it for a moment with an expression of intense fondness; then, raising her eyes to her husband, who stood over her, she laughed for joy.

"'Mother and daughter prospered apace. The little girl became the pet of the house; we all quarrelled for her; but each had to submit in turn. How intelligent! what speaking eyes! what knowing looks! what innocently mischievous ways! mother and child! I wish you could have seen them. I soon marked a striking change: the young Comtesse was now never herself a child. A gentle dignity distinguished her—new-born, it would seem, but natural. I am making my story a long one, but I could talk to you the whole day in this way. So, the months passed on, and the revolution did not abate; and the Comte was sick at heart, and the Comtesse was, as ever, cheerful, contented, happy, and the little one could stand alone by a chair and call out to us all, wherever we were. The Comte, notwithstanding his promise, could not resist his desire to learn more of what was going on than I could inform him of. I seldom went away, for when hawks are abroad it is well to look after the brood; and as I had nothing to gain, and every thing to lose, by venturing out, I thought it best to stay at home. The Comte, on the contrary, was anxious to know every thing. He had made several visits to Calais, first obtaining his wife's consent, although the agony she suffered seemed to fill his heart with remorse; this, however, was soon smothered by his renewed and unconquerable restlessness. One morning he was pleading with her for leave to go again, answering her expressions of fear with the fact that he had been often already without danger. "There is always a first time," said my Agathe, who was in the room. "And there is always a last time, too," said I, happening to enter at that moment. I did not know what they were talking about, and the words came out quite at random. The Comtesse turned pale. "Albert," she said, "content yourself with your Victorine and our babe: go not away from us." The infant was standing by its mother's knee, and, without understanding what was said, she repeated, "Papa—not go." The Comte hesitated: "What a foreboding company—croakers, every one of you —away with such presentiments of evil! Go I will, to show you how foolish you have all been;" and with that he snatched a kiss from his wife and the little one, and started off. The former called to him twice, "Albert, Albert!" and the baby, in imitation, with its little voice said, "Papa, papa!" but the Comte did not hear those precious tones of wife or child, and in a few minutes he was out of sight. I cannot say what was the matter with me; my spirit was troubled; the Comtesse looked so desponding, and Agathe so *triste*, that I knew not what to do with myself. I did nothing for an hour, then I spoke to Agathe: "Wife, I am going across to the town." She said, "Ah, Louis, I almost wish you would go. See how

the Comtesse suffers. I am sure I shall feel easier myself."
Then I told her to say nothing of where I had gone, and
away I went. It did not take me long, for it seemed as if
I ought to hasten. I got into the town, and having walked
along till I came to the Rue de Paris, I was about turning
down it when I saw a small concourse of people on the
opposite corner; I crossed over and beheld the Comte de
Choissy in the custody of four *gens-d'armes*, and surrounded
by a number of "citizens." My first impulse was to rush
to his assistance, but I reflected in time, and contented myself with joining the crowd. One of the soldiers had gone
for a carriage, and the remainder were questioning him;
the Comte, however, would make no reply, except, "You
have me prisoner, I have nothing to say, do what you will."
I waited quietly for an opportunity of showing myself to
him, but he did not look toward me. Presently I said to
the man next me, "Neighbour, you press something too
hard for good fellowship." The Comte started a very little
at the sound of my voice, but he did not immediately look
up. Shortly he raised his head and fixed his eyes on me
for an instant only, and then turned them upon others of
the company with a look as indifferent as if he were a mere
spectator. What a courageous dog! By Heaven, he never
changed an iota, nor showed the slightest possible mark of
recognition; still, I knew well enough he did recognise me,
but I got no sign of it, neither did he look towards me

again. Soon the carriage came up and he was hurried in by the *gens-d'armes*, and off they drove! I made some inquiries and found that the Comte was known, and that they were taking him to Paris.

"'It seems that he had been observed by a spy of the uncle during one of his visits to the town, and although he was not tracked to his home—for he was always very cautious in his movements—yet a strict watch was kept for his next appearance. I went to see the old domestic, but he knew not so much as I. My steps were next turned homeward. What a walk that was for me! How could I enter my house the bearer of such tidings! "*Bon Dieu! ah, bon Dieu,*" I exclaimed, "*ayez pitie!*" and I stopped under a hedge and got down on my knees and said a prayer, and then I began crying like a child. I said my prayer again, and walked slowly on; then I saw the house, and Agathe in the garden, and the Comtesse with the little one standing in the door—looking—looking. I came up—" Albert—where is Albert? where is my husband?" I made no answer. "Tell me," she said, almost fiercely, taking hold of my arm. I opened my mouth and essayed to speak, but although my lips moved I did not get out a syllable. I thought I might whisper it, so I tried to do so, but I could not whisper! The Comtesse shrieked, the child began to cry, and Agathe came running in. "Come with me," said I to my wife; and I went into our chamber and told her the whole, and bid her go to the Comtesse and tell the truth, for I could not. My dear Agathe went out half dead. I sat still in my chamber; presently the door opened, and the Comtesse stood on the threshold. Her eyes were lighted up with fire, her countenance was terribly agitated, her whole frame trembled: "And you are the wretch base enough to let him be carried off to be butchered before your eyes without lifting voice or hand against it, without interposing one word, one look, one thought! Cowardly recreant!" she screamed, and fell back in the arms of my wife in violent convulsions; the infant looked on with wondering eyes, and followed us as we laid the Comtesse on the bed, and then put her little hand on her mother's cheek, and said softly, "Mamma." In a few minutes the Comtesse began to recover. She opened her eyes with an expression of intense pain, gave a glance at Agathe and me, and then observing her child, she took it, and pressed it to her breast and sobbed. Shortly she spoke to me, and oh! with what a mournful voice and look: "Louis, forgive me; I said I knew not what; I was beside myself. You have never merited aught from me but gratitude; will you forgive me?" I cried as if I were a baby. Agathe, too, went on so that I feared she could never be reconciled to the dreadful calamity—for myself, I was well nigh mad. I could but commend the Comtesse to the Great God, and hasten out of her sight. Five wretched and wearisome days were spent.

The character of the Comtesse meantime displayed itself. Instead of sinking under the weight of this sorrowful event, she summoned resolution to endure it. She was devoted to her child; she assumed a cheerful air when caressing it; she even tried to busy herself in her ordinary occupations; but I could not be deceived, I knew the iron had entered her soul. All these heroic signs were only evidences of what she really suffered. Did I not watch her closely? and when the Comtesse, folding her infant to her breast, raised her eyes to Heaven as if in gratitude that it was left to her, I fancied there was an expression which seemed to say, "Why were not *all* taken?" The little one, unconscious of its loss, would talk in intervals about "papa;" and when the mother, pained by the innocent prattle, grew sad of countenance, the child would creep into her lap, and putting its slender fingers upon her eyes, her lips, and over her face, would say, "Am I not good, mamma? I am not naughty; I am good, mamma."

"'Five days were passed in this way; on the morning of the sixth, we were startled by the Comtesse, who, in manifest terror, came to us holding her child, which was screaming as if suffering acute pain : its eyes were bloodshot and gleamed with an unnatural brilliancy, its pulse rapid, and head so hot that it almost burned me to feel of it. Presently it became quiet for a few minutes, but soon the screams were renewed. Alas! what could we do?

Agathe and I tried every thing that occurred to us, but to no purpose: the pains in the head became so intense that the poor thing would shriek as if some one was piercing her with a knife, then she would lay in a lethargy, and again start and scream until exhausted. Not for a moment did the Comtesse allow her darling to be out of her arms. For two days and two nights she neither took rest nor food; absorbed wholly in her child's sufferings, she would not for a moment be diverted from them. Agathe, too, watched night and day. On the third night the child appeared much easier, and the Comtesse bade Agathe go and get some rest. She came and laid down for a little time and at last fell asleep; when she awoke it was daylight; she knocked at the door of the Comtesse—all was still;—she opened it and went in. The Comtesse, exhausted by long watching, had fallen asleep in her chair, with her little girl in her arms. The child had sunk into a dull lethargic state, never to be broken. Alas! Monsieur—alas! the little one was dead! Agathe ran and called me. I came in. What a spectacle! . . . . Which of us should arouse the unhappy Comtesse? or should we disturb her? Were it not better gently to withdraw the dead child and leave the mother to her *repose?* We thought so. I stepped forward, but courage failed me. I did not dare furtively to abstract the precious burden from the jealous arms which even in slumber were clasped tightly around it. Oh! my God! . . . .

While we were standing, the Comtesse opened her eyes: her first motion was to draw the child closer to her heart—then to look at us—then at the little one. She saw the whole. She had endured so much that this last stroke scarcely added to her wretchedness. She allowed me to take the child, and Agathe to conduct her to the couch and assist her upon it. She had held out to the point of absolute exhaustion, and when once she had yielded she was unable to recall her strength. She remained in her bed quite passive, while Agathe nursed her without intermission. I dug a little grave in the garden yonder, and Agathe and I laid the child in it. The mother shed no tears; when from her bed she saw us carry it away she looked mournfully on, and as we went out she whispered, "*Mes beaux jours sont passés.*" Soon the grave was filled up and flowers scattered over it, and we came back to the cottage. As I drew near her room I beheld the Comtesse at the window, supporting herself by a chair, regarding the grave with an earnest longing gaze, which I cannot bear to recall. As I passed, her eye met mine,—such a look of quiet enduring anguish, which combined in one expression a world of untold agonies! Oh! I never could endure a second look like that. I rushed into the house: Agathe was already in. I called to her to come to me, for I could not enter *that* room again. "Wife," I said, "I am going to Paris. Do not say one word. God will protect us. Comfort the Comtesse. Agathe, if I *never* return, remember—it is on a holy errand—adieu." I was off before Agathe could reply. I ran till I came to the main road, there I was forced to sit down and rest. At last I saw a wagoner going forward; part of the way I rode with him, and a part I found a faster conveyance. At night I walked by myself.

" 'I had a cousin in Paris, Maurice Herbois, with whom in old times I had been on companionable terms. He was a smith, and had done well at the trade until the revolution broke out, since then I had heard nothing from him. He was a shrewd fellow, and I thought he would be likely to keep near the top of the wheel. But I had a perilous time after getting into Paris before I could find him. I learned as many of the *canaille* watch-words by heart as I could. I thought they would serve me if I was questioned; but my dangers thickened, until I was at last laid hold of, for not giving satisfactory answers, as *un homme sans aveu*, and was on the point of being conveyed to a *maison d'arret*, when I mentioned the name of Maurice Herbois as a person who could speak in my favour. "What!" said one, "*le Citoyen Herbois?*" "The very same," said I, "and little thanks will you get from him for slandering his cousin with a charge of *incivisme*." There was a general shout at this, and off we hurried to find Maurice. I had answered nothing of whence I came or where I was going, which was the reason I had, at length, got into trouble. I knew Maurice to be a true fellow, revolution or no revolution, and so determined to hold my peace till I should meet him. I found that he had been rapidly advanced by the tide of affairs, which had set him forward whether he would or no. Indeed, Maurice was no insignificant fellow at any rate. The noise of the men who carried me along soon brought him out. I spoke first: "Maurice, my dear cousin, I am glad to find you; but before we can shake hands, you must first certify my—loyalty," I was about to say, but bit my tongue, and got out "*civisme.*" "My friends," said Maurice, "this is my cousin, Louis Herbois, once a valiant soldier, now a brave and incorruptible *citoyen.* He is trustworthy; he comes to visit me; I vouch for him." This was so satisfactory, that we were greeted with huzzas, and then I went in with Maurice. I need not tell you how much passed between us. In short, we talked till our tongues were tired. I found my cousin as I expected, true as a piece of his own steel. He had been carried along, in spite of himself, in the course of revolution, and had become a great man, as the best chance of saving his head. I told him my whole story, and the object of my visit. "A fruitless errand, Louis," said he; "I know the case; and where personal malice is added to the ordinary motive for prosecution, there is no escape. Poor fellow! I wish I could help him; but the uncle, he is in power: ah! there is no help for it." Suddenly a new thought struck him, "Louis, did you come by the Hotel de Ville?" "Yes." "What was going on?" "I looked neither right nor left; I don't know." "Well, what did you hear?" "I heard a cry of *Vive Tallien!* with strange noises and shouts and yells; and somebody said that the National Guards were disbanding, and had forsaken Robespierre; and the people were surrounding the Hotel de Ville." "Then, *Dieu merci,* there is hope. You are in the nick of time; let us out. If Robespierre falls, you may rescue the Comte. He is in the Rue St. Martin; in the same prison is Madame de Fontenay, the *friend* of Tallien, whom Robespierre has incarcerated. The former will proceed thither as soon as Robespierre is disposed of, to free *Madame;* there will be confusion and much tumult. I know the keeper: I must be cautious; but I will discover where the Comte and the lady are secured. Then I will leave you with the jailer; the crisis cannot be delayed another day. Wait till you hear them coming, then shout *Vive Tallien!* run about, dance around like a crazy man—hasten the jailer to release *Madame,* and do *you* manage to rescue the Comte—then be off instantly; don't come here again; strike into the country while the confusion prevails. Come, let us go this minute." And I did go. I found Maurice's introduction potent with the keeper, and, what was better, I found the keeper to be an old companion in arms, who had belonged to the same company with me. We embraced; we were like two brothers; nothing could have happened better. I learned from him all I cared to know. I staid hour after hour; just as I was in despair at the delay, I heard the expected advance. I found my fellow-soldier understood what it meant. I began to shout *Vive Tallien!* as loud as I could cry. In a fit of enthusiasm I snatched the keys from the hands of the keeper, as if to liberate the lady, while my comrade opened the doors to the company. I hied first to the Comte's room. In one instant the door was unlocked. "Quick!" I whispered; "follow me—do as I do. Shout, huzza; jump this way and that—but stick close to me." In another minute I had unbolted the door of Madame de Fontenay, making as much noise as I could get from my lungs—the Comte keeping very good time to my music. So, while we were shouting *Vive Tallien!* at the top of our voices, Tallien himself rushed in with a large party. I took the opportunity to gain the street, and, without so much as thanking my comrade for his attentions, I glided into an unfrequented lane, the Comte at my heels; and I did not stop, nor look around, nor speak, till I found myself under cover of an old windmill near St. Denis, where I used to play when I was a boy. There I came to a halt, and seizing the Comte in my arms, I embraced him a thousand times. I took some provisions from my pouch, which my cousin had provided, and bade him eat, for we should stand in need of food. We then proceeded, avoiding the main road, and getting a ride whenever we could, but never wasting a moment—not a moment. I told the Comte what had happened, and that he must hasten if he would see his wife alive. At last we came near our house. The Comte could scarcely contain himself: he ran before me; I could not keep up with him. How my heart was filled with foreboding! how I dreaded to come nearer!—but apprehension was soon at an end. There was my little cottage, and in the doorway, leaning for support against the side, stood the Comtesse, gazing on vacancy—the picture of despair and desolation. At the sight of her husband, she threw out her hands and tried to advance: she was too feeble, and would have fallen had he not the same moment folded her in his arms.

"'*Bien, Monsieur!*' continued Louis Herbois, after clearing his voice, 'the worst of the story is told. The Comtesse was gradually restored to health, and the Comte was content to remain quietly with us till the storm swept past; but the lady never recovered the bright spirits which she before displayed, and the Comte himself could never speak of the little one whom he kissed for the last time on that fatal morning, without the deepest emotion. It seems to have been destined that this should be their only affliction. The uncle was beheaded in one of the sudden changes of parties the succeeding year, and in due time the Comte regained his estates. Sons and daughters were born to them, and their family have grown up in unbroken numbers. The Comte and Comtesse can scarcely yet be called old, their health and vigor remain, and they enjoy still those blessings which a kind Providence is pleased to bestow on the most favoured. But the Comtesse de Choissy will never forget the child which lies *there*. Twice a year, accompanied by the Comte, she visits the cottage. She lays with her own hands fresh flowers over the little grave, and waters the moss which overspreads it; and the tears stand in her eyes when she looks upon the spot where we buried her *first-born*. We have engaged that every morning we will renew the flowers, and preserve the mosses always green. It is a holy office, consecrated by holy feelings. Ah! life is a strange business: we may not be always serious, we cannot be always gay. God grant, Monsieur, that in Heaven we may all be happy!'

"I have given you the whole story," said Mr. Belcher, after a short pause; "but look, the sun is out; let us go to the Courtgain."

. . . .
. . . . . .

I retired that night with a great many new impressions. My head was full of Mr. Philip Belcher, and before I went to sleep I took a dozen different surveys of his character. There was a genuineness in it which was positively charming. To be sure, he was a little too abrupt—a little too positive; but that I could excuse in one so many years my senior. I began to turn over in my mind his theory of travel, and this led me into a sort of review of my own plans. We were going to Paris to pursue a distinct course of study, walk the hospitals, and attend lectures— that was the object of our coming abroad. But along with this, if the truth must be told, at least so far as I was concerned, floated agreeable visions of the brilliant capital of France, which were not associated with the hospital or the lecture-room—visions somewhat indistinct, but conveying ideas of novelty, gayety, pleasure: embracing, perhaps, sight-seeing, lion-hunting, promenades on the Boulevards, visits to the cafés, evenings at the opera or the saloons, ascending columns, traversing gardens, lounging in shops, witnessing the ascent of balloons, attending executions at the Place de Grêve, and so forth; from all which were to spring innumerable charming adventures. And here my visions would fade quite into air, and throw me back on vague conjecture. I admit there was nothing extraordinary in this—really nothing at all. To an American youth, brought up in all the strictness of a New England education,—to which I beg leave here to give my hearty commendation,—there can be no greater change than to transport him suddenly to Paris. The experiment is a hazardous one. The chances, certainly, are two to one that he goes to the devil, or, if he stops short of that destination, it is with so much injury to character and person that it takes the greater part of his life to repair damages. But I am digressing from my subject, which was Mr. Philip Belcher and his peculiarities, and the effect our intercourse was likely to have on me. It had already set me thinking in a new direction. For I had really formed no distinct purpose of understanding the French character, content to assume the commonplace and ridiculous notions usually entertained on the subject by those who speak the English tongue. Besides, I had not a thought about any part of France, except Paris. It had never occurred to me that the space between Calais and Paris was inhabited—that it contained human beings. Now I did think of it. A sudden desire arose to know something about them. How did they manage their every-day affairs; what were their habits, their social customs; how did they live, marry, and die; what were the sports of the children, the occupations of the young men and women, the employments of the old; and so revolving a thousand things newly sprung up in my brain, I fell asleep.

CHAPTER II.

A SURPRISE.

"Upon my word, you have taken rapidly to French habits. Ten o'clock—almost, and you sound asleep."

I opened my eyes, and saw Partridge standing over me with a most amused expression of countenance. "My dear fellow," I exclaimed, starting up and seizing his hand, "I am so delighted to see you. Where have you been, and what have you been doing all this time, and—and when did you arrive?"

"I will tell you the whole story one of these days; I got in last night after you were in bed, and was coming straight to your room, but an odd old fellow, with very square-toed boots, who took the liberty of claiming me as a countryman, when I entered the house, said he thought I had better not disturb you. One would suppose he had known you a thousand years: he seems to take quite an interest in you. Has he an unmarried daughter?"

"It must be Mr. Belcher."

"Who the deuce is Mr. Belcher? But never mind that

now. You must hurry and dress. I have taken seats for both of us in the diligence; it starts in just one hour. Only think of it — to-morrow night we are in Paris! Bravo! huzza! tira lara! Come, come—en avant. 'The climate's delicate; the air most sweet;' and we set off in just fifty-seven minutes."

A bucket of cold water seemed to fall with one dash upon my newly developed ideas of travel and wayside investigation. Partridge discerned something in my appearance, for he exclaimed, "What is the matter? any thing wrong?"

"Oh, nothing at all; but don't you think you are a little hasty in your decision to push on so rapidly to Paris?"

"*Rapidly!*" cried my friend; "I am sure *you* need not complain. Why, I have been pitying you the whole morning for what you must have suffered the last five days in this miserable hole. I have been up since five o'clock, and have seen all I desire to see of this place: I want to get out of it."

"But," I interposed, "suppose we proceed leisurely over the road, and see something of the inhabitants between here and Paris, and learn their manners and customs, that we may really know something about the French people."

"I will lay a hundred to one that your Mr. Belcher has been stuffing your head with this nonsense; indeed, the old chap I think had some design on me, but I gave him the slip. Now, seriously, what do you want of these folks by the way? You will see coarse peasants; men and women with wooden shoes, or with no shoes, at work in the fields, or driving or riding a donkey small enough for me to put in my pocket; people who live on bread, soup *maigre* and carrots, and who do not possess the first point of interest for any one. Besides, what a miserable plan for our becoming perfect in French. I thought I could speak the language like a native, and I find I can neither understand what is said to me, nor make myself understood in return."

I drew a long sigh as I perceived my late visions of travel-life fade away, and replied almost reluctantly, "I suppose we must go, especially as you have engaged seats"—"and paid for them," interrupted Partridge; "and more than that, I have paid a little Frenchman for giving up his seat in the *interieur* and going into the *rotonde*, and now we shall be seated *vis-a-vis* to two fine-looking women. I have seen them both, and I calculate on improving my French vastly on the journey."

There was nothing more to be said, so I hurried down, breakfasted, we mounted into the *interieur*, (the ladies were there before us,) and away we rattled. No: we did not rattle away; for just as we were about to do so, as I thought, the *chef du bureau*, or, as we would say, the "stage agent," comes to the door and proceeds to examine us by his list.

"Numero 1, Madame Le Preux. Numero 2, Monsieur Taige."

"That means me," cried Partridge,—"*Ici, Monsieur;* now I know the French for my name."

"Numero 3, Madame Vigny. Numero 4, Monsieur Taige *encore.*"

"That's you," said my friend, addressing me; "I put the two seats down in my name; all right, *Monsieur*, that is, *c'est juste.*"

"Numero 5, Monsieur Le Preux. Numero 6, Capitaine Duclos."

But the seat was vacant, and no Capitaine Duclos responded. This led to a slow, careful, scrutinizing call of the six numbers and the six names over again, as if in some way it would come out right on another trial. So these were repeated, with considerable pause between each. Although "Capitaine Duclos" was, when reached, pronounced with a desperate emphasis, the "Capitaine" did not answer, and the seat remained vacant. A Frenchman when he is puzzled wears a singular expression of visage; and, strange as it may be to an American, here was matter to puzzle any *chef du bureau* of any of the messageries in France. If Capitaine Duclos had enrolled his name and paid for his seat, why was not Capitaine Duclos on the spot?

It was a question that could not be answered, and the *chef du bureau* had a right to be puzzled; still he did not relax his efforts. He first made the tour of the entire vehicle; examined the *rotonde*, the *coupé*, and climbed up to the *banquette;* then he came back to us and made a silent count. Finally, he caused proclamation to be made over the court-yard—" Numero 6, Capitaine Duclos."

It was all of no avail.

Our ladies began to grow impatient. They expressed themselves very decidedly too. " Why should they be obliged to wait for Capitaine Duclos? It was ten minutes past the hour—the other diligence had started—why did not the *conducteur* proceed? Such conduct was without parallel in the world's history," and so on. Partridge and myself chimed in with assenting words and gestures, and thus an acquaintance was established at the start, or, I should say, before the start. In the midst, however, of our energetic remarks and indignant remonstrances, Capitaine Duclos appeared. He did not, by any means, approach with the air of a man who has been the unfortunate cause of inconvenience to an entire company : he was not heated, or excited, or even in haste, nor was there any thing deprecatory in his appearance. He walked with remarkable self-possession to the door of the *interieur*, coolly took an observation to ascertain which was his seat, then with a great deal of deliberation selected a place for his sword, which was carefully incased; next followed a small leathern box, and after that an old military cloak. Then, first glancing around lest he should seem at all in a hurry, he slowly seated himself, while the *chef* proceeded once more, but this time silently, to count our company. There was no difficulty now. "Numero 6" was all right; so with a complacent nod to the *conducteur*, which indicated that the arrangements were complete, the diligence was allowed to get under way.

The indignation of our company against Capitaine Duclos speedily subsided as we were galloped out of the town, although Partridge occasionally eyed the imperturbable *militaire* with something of a defying manner. I am not, however, about to make a hero of the Capitaine. I have given an account of his first appearance because it presented to me a phase of French character which I had never before observed. He exhibited nothing of what is ordinarily called "French politeness," although he was not in any degree rude or uncivil: neither was he reserved after the fashion of an Englishman: but there was something in his *tout ensemble* which said, "An officer of the army of the *Grande Nation* can have no apology to offer for keeping a diligence waiting ten minutes;" and what is more, the *chef du bureau* appeared to be of the same opinion.

But enough of our military friend. To us every thing

was novel, and, of course, every thing was delightful. The very vehicle in which we were transported was a curiosity; while those little peculiarities, so insignificant in themselves, yet so striking to the stranger, and which it is hardly possible to enumerate, kept our minds on the alert with pleasing excitement. The rain had settled the dust, and the verdure of every kind was tinged with a deeper green. The ladies who were passengers with us were very sociable, communicative, and entertaining; freely correcting our blunders, and willingly answering the many questions we crowded upon them. At length, the country assumed a more pleasing aspect. We passed beautiful little hamlets, half hid with clustering shade-trees, in each of which, with its ponderous bell suspended in the tower, was an old stone chapel where the inhabitants came to worship. How much I wanted to leave the diligence and walk through these scenes! The graveyards, filled with strange-looking black crosses and fantastic ornaments, produced another class of impressions, which were speedily changed for others still. Women, with immense straw hats, were labouring in the fields, or going to or returning from them. Images of the Holy Virgin could be seen placed in small niches in many of the houses, and tastefully decorated. As we passed through the larger villages we amused ourselves reading the signs over the different hotels. I can still remember many of them—so vivid are first impressions—such as "*Hotel de les Syrenes*"—"*Hotel de la Maison Blanche*"—"*Hotel du Cheval D'or*"—"*Hotel du Sauvage*"—"*Hotel des Gentilhommes*," and so forth. I recollect the two last struck us as an example of the meeting of extremes. The donkeys we encountered by the way formed no trifling object of attention; and it seemed the farther we advanced the smaller they became—and that the smaller they became, the heavier they were laden. What immense panniers, and what very little donkeys! We had great amusement at dinner in fighting our way through the varieties of a French *cuisine*, and perceived very little difference in the scramble after food by passengers from a diligence and those from a stage-coach at home.

We came back to the diligence in fine spirits. Dinner had put us on most excellent terms with ourselves, and in good humour with every body.

> "From school to Cam or Isis, and thence home;
> And thence with all convenient speed to Rome,"

repeated Partridge, gayly.

"Go on," I cried, while I continued the quotation for him—

> "With memorandum-book for every town,
> And every post, and when the chaise broke down."

—Here Partridge again took up the verse—

> "His stock, a few French phrases got by heart,
> With much to learn, but nothing to impart."

"You omit the best of it," said I, as my friend suddenly stopped—

> "Surprised at all they meet, the gosling pair,
> With awkward gait, stretched neck, and silly stare,
> Discover huge cathedrals built with stone,
> And steeples towering high—much like our own!"

And now Partridge insisted I should stop, declaring we had satirized ourselves sufficiently without my finishing the quotation. Then one of the ladies inquired what we had been saying; so we attempted a translation of the lines, first one taking the lead, the other correcting, and then vice versa, until we got the whole company laughing. In short, the utmost hilarity prevailed for the remainder of the day, and we rumbled into Amiens much more favourably impressed with France, Frenchmen, and Frenchwomen, than when we left Calais that same morning.

Great was our progress in *lingua Franca* that evening at the hotel,—and in the morning too,—for innumerable were our demands upon man, woman, and child, in and about the house, and marvellous the alacrity manifested in answering them.

All our company were, it seemed, bound for Paris; for all were ready on the starting of the "vehicle," as Partridge called it, and took their seats precisely as they did on the previous day: the door was just closing, when

Madame opposite us, who had been looking over her bill, suddenly exclaimed, "*Tenez, tenez un moment!*" and, darting out, to the astonishment of every body, she disappeared in the hotel. Presently she came running back, having in her hands two wax candles, which she bore away in triumph, exclaiming, "*Mes bougies!*" whereupon Partridge seemed possessed with a sudden idea, for he, too, rushed from the carriage, cleared the court-yard almost at a bound, sprang up the broad staircase, and in an instant came down again, also with two wax candles, which he exhibited to *Madame* with a knowing nod, to which the lady replied with a sympathetic assent. The whole affair was Greek to me, which, however, I readily comprehended when Partridge exhibited our bill, in which I read, among other items,

<blockquote>2 bougies . . . 2 francs.</blockquote>

"Shameful—outrageous—not to be tolerated!" exclaimed our fair friend.

"Perfectly so," responded Partridge, seriously.

"*Voila, Monsieur,*" she continued, showing the ends of the bougies, "our candles were not lighted ten minutes, and to be charged a franc for each! Heaven knows I don't want them"—here she wrapped them very carefully in some paper and put them into her bag—"but it was so monstrous that I would not endure it. Had it been half-a-franc each, it would have been quite another thing."

"Madame is perfectly right," replied Partridge, "I quite agree with her; 'tis really a monstrous piece of work, and, as you say, little as I require the article"—and he thrust one without wrapper into either pocket—"I could not but follow your example." Thereupon, with a look of serious dignity, my friend settled himself once more in his seat.

Without further detail, I will say that the second day of our journey was even more agreeably passed than the first. Full of excitement—and *such* excitement too, so buoyant, so ample, so unmixed, as only the young experience—we rattled over the heavy pavements of Paris—Paris, city of pleasures and of scientific pursuits, distinguished for frivolity, distinguished for research. Extremes meet once more. The *Hotel Sauvage* and the *Hotel des Gentilhommes* again. But I had no time to moralize. The bright gas-lights from the various shops and cafés flashed brilliantly as we passed along, while the continual crack of our postillion's whip seemed to invest the diligence and all its passengers with an additional importance. We dashed along the *rue St. Honoré* and turned into the *rue de Grenelle* and brought up at the general office of the then *messagerie Royal*. A large hotel flanked one side of the court, and as the evening was advanced, for want of a better place for *logement*, we went there. The charm of novelty made us regard every thing as possessing a "pleased aspect." The morrow brought some drawbacks. We were astounded to find the streets for the most part without sidewalks, while the gutter (like the trickling gore from the extinguished eye of Polyphemus)—

"Luminis effossi fluidum———"

laid its dirty course through the middle of the street. Several unmentionable nuisances offended our fastidious sense, and led Partridge to purchase a record-book for the entry of the particular cases which came under his observation in *la belle Paris*. But these were only trifling *impedimenta* to the ardent zeal with which we prosecuted our discoveries. But two or three days were spent rambling over the finer part of the town, before we went across the Seine to examine the *quartier*, which was to be our residence for some time to come. If we had cause for complaint before, what were we to say now about streets, *trottoirs*, gutters, and cleanliness, or rather filth, generally. It was well, after all, that we were struck with the worst at first, and had room afterwards to become interested with what had escaped our observation. Partridge had an acquaintance, a young medical student, from the state of Vermont, who lodged in the *rue Copeau*. Thither we traced our steps, and finding the number, were directed how far to mount, and we ascended accordingly. We found Vincent (that was his name) in his room, in which were gathered some eight or ten students, who were talking and laughing, discussing politics—Alibaud had just shot at and missed Louis Philippe—smoking, &c., &c., &c. Our countryman received us most cordially, introduced us formally to every one of the company, which evidently composed a set, said it was very fortunate; two capital rooms were vacated that very morning which would suit us exactly; if we chose he would step at once and tell Monsieur Battz that they were taken. "Bravo!" shouted several, "our numbers will be kept good; that is, if we can get Etienne out of the *conciergerie*." I learned afterwards that he had been arrested in that house a few days before, and sent to prison, for some extra revolutionary manifestations.

I think Partridge and I both *felt* like hesitating, but we did not hesitate. It would have been rude to Vincent; besides, as absolute strangers, we could not do better than accept "any port" for the present, until we could see and judge for ourselves. So we thanked our friend and assented to his suggestions, and in three minutes Monsieur Battz appeared, followed by the two Mademoiselles Battz, by all three of whom, with Vincent for an escort, we were ushered into our apartments, which were really delightfully situated in the rear of the house *au troisième*, and overlooking a very pleasant garden.

"I have told my friends, Monsieur Battz," said Vincent, "that they are to have the rooms at the same price which Rolles paid for them."

"*Pardon*, Monsieur Vincent, but really *c'est impossible*, you know——"

"Otherwise I must take them across the way," continued Vincent, coolly, "you see half the rooms over there are to let, and as to price——"

"*C'est marché donné*," interrupted Monsieur Battz, hastily. "*N'importe, j'y consens, mais j'y perds, en vérité.*"

"I know it, I know it, Monsieur Battz," replied Vincent, "I know you lose money by it; but then you are such a benevolent man, Monsieur Battz, and, Mademoiselles, I am sure *you* wish my friends to take the rooms?"

"*Oh, certainement, Monsieur, assurement,*" said both the young ladies, with a graceful bow and a score of French protestations.

"Seven sous per bottle for red wine," continued Vincent to us.

"Ten, *Monsieur*, ten sous," once more interrupted Monsieur Battz; "ten, I repeat, Monsieur Vincent, for all, except my old lodgers——"

"Not including the bottle," proceeded Vincent, without

heeding the interruption, "but it is of course understood that no one keeps his wine into the second day."

I felt almost sorry for poor Monsieur Battz, who seemed broken-hearted at Vincent's peremptory way of disposing of matters, to say nothing of the pantomime kept up by the young ladies with deprecatory shrugs and most expressive grimace, as if they had said—

"You see, Monsieur, we are actually submitting to a great loss in order to oblige you."

I was on the point of doubling the price of rooms, wines, segars, and every thing else, so much was I affected by these manifestations, when it occurred to me that Vincent very naturally knew much better than I about the matter, so I held my peace.

Going back to his room, we soon engaged in general conversation with different members of the company.

"Ah, Partridge," cried Vincent, "I knew you would follow soon—I saw it in your countenance the day I said good-bye to you in New York.

'Home-keeping youth have ever homely wits.'

And now you will admit it is something better

'To see the wonders of the world abroad,
Than living dully sluggardized at home.'"

"I don't know any thing about the world," said Partridge, demurely. "I came to Paris to study medicine."

"Oh, of course, of course, to study medicine, certainly," said the other, laughing, "but then you know one may——"

Here some of the party interrupted him, by saying it was time to go to the lecture, when there was a general rising, and the company set off for the lecture-room, while we remained behind, in order to get comfortably settled in our new rooms. The house was an antique building of stone, erected with considerable architectural taste. It was very large and spacious, and had, no doubt, at an earlier period, been the hotel of some distinguished person. It was used now entirely for students' apartments, with a *salle-à-manger* below and billiard-room adjoining, and a wine-house close at hand.

The next day we commenced to "walk" the Hospital *de Notre Dame de Pitie*, and were soon fairly launched on our new life.

I may hereafter speak of the different members of our company at the *rue Copeau*. It will be best to introduce them as occasion may require.

There was one young fellow in particular, to whom I took a great fancy. He was an Englishman, by the name of Clements, a quiet, civil person, not altogether reserved, but possessing a species of unobtrusiveness, which usually conceals much that is worth seeking out and cultivating. We became intimate. He had been several years abroad, and pursued medicine with an enthusiasm curious to witness. His means were ample, and it was the mere love of his profession which kept him in Paris. He was, besides, a thoroughly educated and accomplished scholar. From the moment of our arrival, Clements seemed disposed to do all he could for us. He took pains to inform us about every thing desirable to be known by new-comers, and gave us many valuable hints as to our hospital course, lectures, and so forth. We used to take many walks together over the city, with which Clements' long residence in Paris had made him familiar. We indulged, too, in frequent discussions together, on a great variety of subjects; sometimes they assumed a serious phase, sometimes they were lively, sometimes sentimental, sometimes matter-of-fact, but to me always agreeable.

Clements had one peculiarity. He was a very strong believer in the good faith of his own sex. He could not read any ordinary work of fiction, or even the merest story, (for almost all tales relate in some way to love affairs,) without losing his patience.

As I stepped into his room one afternoon I found him just closing a small volume, which, as I entered, he tossed out of the open window with an impatient gesture.

"How perfectly disgusted I am," he cried, "with this absurd, sickening, lackadaisical cant, which is for ever crying up the wrongs and silent endurance of injured woman, and the inconstancy and selfishness of 'tyrant man.' By the way," he proceeded, "do you know there is a class of romance writers and poets, among whom, I am sorry to say, are some distinguished names, who invariably use for a 'stock in trade' such profound watch-words as the following: 'With man, love is a pastime; with woman, her very existence.'—'Man gives to woman his leisure; woman gives to man her life.'—'Man is inconstant; woman is true.' When I hear such apothegms daily repeated, and the changes rung upon them over and over again, (all this being predicated of man because he is man, and of woman because she is woman,) I am ready to exclaim, with the clear-hearted Burchell, 'Fudge!'"

"I think you are a little too sweeping in your denunciation," I replied. "There does seem to me to be a constitutional difference in the sex; perhaps, however, it is but the result of education; but for the matter of watch-words, as you call them, what have you to say to the lines of the great dramatist?—

'———————Were man
But constant, he were perfect: that one error
Fills him with faults; makes him run
Through all the sins:
Inconstancy falls off, ere it begins.'"

"I have to say," answered Clements, "that if you

substitute 'woman' for 'man' in the passage you have recited, it would contain just as much truth, but not so much poetry, as it now does. For the same reason, I hold that—'Frailty, thy name is man,' is just as true as—'Frailty, thy name is woman.' No, my dear fellow," continued Clements, "you cannot reason me out of the belief, that the Deity made man as true of heart, as earnest in his love, as devoted in his attachment, as woman. The Scripture records that 'in the image of God created he *him;* male and female created he *them;*' and surely that work must have been well done which God himself pronounced 'very good!' That man has more to occupy and distract his attention; that he is, in a majority of cases, continually engaged in a struggle with need, and, in consequence, that his affections are less seldom fixed than those of woman, is true enough. On the contrary, the life of woman, as society is constituted, is calculated to give to her impulses a hot-house growth, (I say nothing of the direction;) so that love with her becomes neither a healthful passion nor a refined friendship, but simply a feverish longing, derived from that strange heart-vacancy which every young girl, after reaching a certain age, is sure to experience. If at this period some natural and agreeable occupation could be provided, which should serve to keep both the mind and the heart in a healthful tone; if man could be less engrossed with cares and woman less with—nothing, I believe broken hearts would be nearly equally divided between the two sexes. But let us have done with the subject. Few people agree with me, and I am myself so stubborn that I agree with but few; so let us take a walk."

I assented, and we took our course over by the Pantheon and wandered into the *rue d'Enfer*. At length my companion stopped before one of the houses, and, pointing to a window, said:

"I lodged in those rooms for a long time. It was a singular affair drove me away from them. I have never related it to any human being, and for very good reasons. I have half a mind to tell you the story—I know I can trust to your discretion—for it comes very *à propos* of our conversation before we started out, and presents a case of devotion on the part of man worthy of record. I was an eye-witness of what I am about to tell you, and——but the street is no place for my story; here is a quiet spot where I used often to go. You shall have the narration with some of Antoine's best *café*." Clements led the way into a neat apartment, and selecting a retired corner, gave his order to the *garçon*, which being supplied, he prepared his coffee, sipped a spoonful of the beverage, cast a glance over the room, and commenced as follows:

CHAPTER III.

THE STORY OF LUDWIG BERNHARDI.

I FIRST came to Paris four years ago to attend medical lectures. The revolution which made Louis Philippe king of the French had subsided. The city was quiet, except when disturbed by occasional plots against the king's life, manifested by the letting off of pistols, blunderbusses, and "infernal machines," in a way that none but Frenchmen know how to appreciate.

There were at that time in Paris an unusual number of students; I suppose between twenty and twenty-five thousand. These were made up, as you very well know, from almost every country upon the face of the globe. Nearly all of them had apartments "*sur l'autre côté du Seine*," in the part denominated "The Students' Quarter," quite as you now see them. By the way, although we form here, in a measure, a community of our own, still you must not suppose it is similar to a community of German students: far from it. For while the size and immense resources of Paris present continual and varied temptations for the idler and the pleasure-seeker, and the excitement of politics (your student is always a true republican) gives a zest to the life even of the most studious, they serve at the same time to break down that barrier which always stands, as an absolute division, between the students in German universities and the "outside" world. Therefore in Paris there is more of refined debauchery; in the universities, more out-and-out, dare-devil dissipation and hardihood: in Paris, numerous intrigues, an occasional assassination, and few duels; in the universities, less intrigue, no assassinations, and half-a-dozen duels *per diem*. I need not tell you who are now living on the spot, that the morals of the students generally were bad—deplorably bad. With comparatively few exceptions, each student lived with his *maîtresse*, who, besides being his faithful and attached "friend," (I use the parlance of the town,) performed the part of his housekeeper, saw to the preparation of his *café*, and looked kindly after his wardrobe. These alliances, you know, sometimes continue for years, with fidelity upon both sides. But it is not my purpose to go into any detail of what has so often been spoken of: I only allude to it now, to make my story intelligible.

My lodgings were here in the *rue d'Enfer*, at the spot I pointed out to you; several acquaintances had apartments in the same place. Most of us attended upon the same lecturers and walked the same hospitals.

Directly across the street stood an antiquated—even for the *rue d'Enfer*—stone house, on which I had never seen placarded "*Apartemens à louer*," but where lived a pale, slender, sad-looking, light-haired young man, who came forth daily and proceeded to the lecture-room or to the hospital. As he happened to make similar rounds with myself, I soon got acquainted with him; that is, we spoke when we met, walked along together if we fell in company, and conversed, though sparingly, on ordinary topics: further than this, however, I found it hard to push my new acquaintance. He was a native of Wirtemberg, and his name was Ludwig Bernhardi. There was a mystery about him which I could not fathom. His manner was neither cold nor distant, but beyond a certain point no one could get with him. He declined every invitation to visit, and never invited any one to visit him. He kept very quiet, went to no place of amusement, and never mingled among the students. There was a large garden attached to the old stone house where Bernhardi lodged, and a lively young Frenchman, of our company, one day ran through the hall and looked out into this garden, where he saw, as he declared, the pale student walking with a beautiful young girl. After this announcement the mystery for a time was cleared up: "Bernhardi was so engrossed with his '*chère amie*' that for the present he cared for nothing better;" "The Wirtemberger was no fool, after all;" "The German was silent and shrewd;" and so on and so forth. For myself I did not fall in with these generally-received explanations. There was something about that pale and saddened face, that suffering and subdued air, which was inconsistent with any of them; at least, they did not satisfy me. No one had as yet got a glimpse of the fair maiden, except the young Frenchman, and he made his companions half crazy with his descriptions of her beauty. After a while curiosity began to prevail again. Singular to say, the girl was never seen to come to the street, either by herself or in company with her lover. Now Bernhardi might have lodged a dozen nymphs in the old stone house, and not a soul would have taken notice of it so long as things had gone on after a natural way; but when the student never walked out with his sweetheart, never took her to the "theatre," nor to the "gardens," nor to a "spectacle;" when the maid never appeared at the window, nor in the hall, nor at the little fruit-market, where ripe cherries and strawberries, the usual accompaniments of a student's breakfast, were procured by the devoted "friend;" when, to crown the mystery, the young girl was observed one day to come to the street-door, and was about passing out, while

Bernhardi hurried after her, and, partly by force, partly by entreaty, urged her away; the curiosity of every one was excited, and the matter became a topic of general conversation and remark. Notwithstanding all this, no person, that I am aware of, said aught to the student on the subject. He was an individual that no one would care to take such a liberty with. One could not but entertain a vague apprehension that by so doing one might rouse a sleeping devil which should not be so easy to lay.

About that time a new-comer took possession of an apartment in our house which had been vacated a few days previous. He was from Marseilles; a tall, swarthy, black-looking creature, brawny and muscular, a savage in appearance, with a reckless, swaggering gait, a bullying air, a fierce, impudent mien. He was just the sort of fellow to domineer over the timid and the yielding, and to hide his crest in presence of true courage and resolution. To persons of such description I generally give a "wide berth:" I would rather avoid than quarrel with them. There are no laurels to be gained in silencing a barking dog; and there is something humiliating in a conquest over a poltroon and a coward.

For this reason, I made it a point to have as little to do with Balaiguer (that was the name of the Marseillese) as possible. Some of my comrades were particularly taken by his bold front and egregious pretensions; and with a certain class he got to be both leader and oracle. I soon discovered him to be an infamous creature. He was, besides, a miserable debauchee, and was actually doing serious injury to habits and morals among a class where habits and morals were in all conscience lax enough.

Balaiguer was not long in getting hold of the story of Bernhardi. Then he swore a vulgar oath that " he would unearth this sly fellow; he would see whether a man had a right to keep his pretty mistress shut up in a cage like a bird. He would pay the minx a visit, and what was more, by ——! he would carry her off, *nolens volens*, before the little Dutchman's face and eyes."

I happened to be present at this harangue, which was made one day to a knot of students assembled in the "*salle-à-manger*." Balaiguer's announcement made me shudder; not that I feared for the safety of the parties threatened; but a presentiment suddenly came across me that *death* would be in the mess which the Marseilles was brewing.

The next day Majendie was to lecture at eleven upon the "cause of pulsation." I had returned from my usual morning visit to the hospital, where we had the privilege, as you now have, of "following" Louis, and was quietly seated at my little breakfast-table, when, after a light knock, the young Frenchman, who had reconnoitred the garden across the street, entered the room. I should have mentioned that he was a Parisian, of good family, and although gay, thoughtless, and fond of a frolic, had nevertheless a nice sense of honour, coupled with real refinement of character.

"Do you know," said he, "that I feel reproached about our neighbour opposite? Here is Balaiguer, who swears that as soon as Bernhardi goes to the lecture he will run over and make love to his mistress: now I know the *bête* will do her some violence, and it is all owing to the foolish stories I have told of my seeing her in the garden; I thought but to have some fun with my comrades; to tell you the truth, the girl was beautiful, but there was something in the looks of both that has made my heart ache ever since. Believe me, it is not as we suppose; and yet my jokes have set on this *coquin*. What shall I do?"

"You are a noble fellow," I exclaimed, involuntarily. The young Frenchman took my hand and pressed it to his heart. The impulsive words were appreciated. "We will step at once," said I, "to Balaiguer. He must not think of such a thing. We do not want to quarrel with him; but we———"

"Fear nobody," interrupted the young Frenchman. "Let us go."

Accordingly, we proceeded to the apartment of the

Marseillese. It wanted but ten minutes to eleven. If I made any delay I should lose even a tolerable seat in the lecture-room, so I came at once to the point. Under other circumstances I might have been less direct. "Balaiguer," said I, "our friend here informs me that we are altogether on the wrong scent as to Bernhardi, and that there is nothing over the way to excite your curiosity or repay your gallantry. We hope, therefore, you will let our neighbour rest in peace."

"Bah!" said Balaiguer; at the same time putting the forefinger of his right hand under his eye, and pulling down the lower lid, he exclaimed, in a jeering tone, "*à d'autres!*"

"I suppose I understand you," I continued. "Now look you, Monsieur Balaiguer, we students love fair play. I am no informer, but I give you notice that I shall warn Bernhardi of what you would be at. Good-morning."

"You could not do me a greater favour," shouted the Marseillese, as the young Frenchman and I passed from the room. "Tell the Dutchman to hurry, for I shall make short work of it."

We descended to the street, hoping to see Bernhardi as he came from his room; we were too late. Our *concierge* informed us that he saw Monsieur leave his house nearly five minutes before we came down. "Hasten after him," said the young Frenchman. "I will not go to the lecture; I will remain in my room. *Mon Dieu!* I am quite nervous."

I had nearly half-a-mile to walk, or rather to run, for I believe I ran all the way. As I anticipated, the room was crowded. The lecture had commenced, for Majendie was punctual, and he had much ground to go over. A goose, which was to be dissected alive, in the course of his remarks, stood upon the table, in charge of a favourite student, and as I entered the familiar "*comprenez-vous*" of the lecturer fell upon my ear. I heard nothing more. I glanced anxiously up and down, over and across the room, but could not see the object of my search.

"What the devil is the matter with you?" said my friend D——, taking hold of me.

"Nothing; I want to find Bernhardi."

"There he is, away in that corner. Don't you see him?"

I took a direct course for the corner, sometimes over a student's back, sometimes over the benches, and laid my hand upon his shoulder. "You had better go home!" I whispered in his ear.

Swift as thought the German sprung to his feet. His face became livid; his eyes started from their sockets.

"Quick!" said I.

Bernhardi had disappeared.

I do not know how I sat out the lecture. I have

some recollection of seeing the poor goose struggle, or try to struggle, and of the complacent air of the lecturer, as he mingled his "*Entendez vous?*" "*Eh bien! voyez vous?*" with the cries of the suffering creature, while he deliberately cut away muscle, and nerve, and tendon, in the gradual illustration of his subject. But my thoughts were elsewhere. I saw in my mind Bernhardi and the Marseillese. I pictured every conceivable catastrophe; and so engrossed did I become in this, that the first hint I had of the completion of the lecture was the general uproar consequent on clearing the hall. I hurried out by myself, and hastened to the *rue d'Enfer*.

Going up the staircase I saw a few drops of blood scattered along. At that moment the young Frenchman opened the door of his room, and drew me into it. His mirthful countenance at once relieved me.

"Come in—come in!" he exclaimed; "I have been watching for you. Balaiguer has caught it;" and he began laughing immoderately.

"Don't laugh any more, for Heaven's sake, till I know what it is at!"

Whereupon, in few words, the young Frenchman informed me that very soon after I left Balaiguer crossed over to Bernhardi's quarters; that he stationed himself at an open window to watch the other's movements; that after the lapse of some five minutes he heard a violent scream, and was about running across to protect the party assailed, when Bernhardi came tearing down the street like a madman, and rushed into the house and up the stairs, and in less than a minute the Marseillese was seen rolling from the top to the bottom; that he picked himself up and skulked back into his room, bleeding, but, as my companion feared, not much hurt.

After expressing our mutual delight at the termination of the affair, I went to my own room. I took it for granted that the matter was ended, for I knew that Balaiguer had not courage to push it further, and I supposed that Bernhardi would rest satisfied with the chastisement he had already inflicted. I was mistaken; for in a few minutes a knock was heard at my door, and Bernhardi entered. He was pale as death; his eyes glistened with intense hate and desperation; his soul appeared harrowed by the most violent emotions; but when he spoke, his words fell slowly, and were articulated naturally.

"I am under an obligation to you: for that reason I come here. I would be still deeper in your debt. Will you go for me to the *wretch* and demand immediate satisfaction? I say *immediate!*"

"Are you not carrying the matter too far?" said I, soothingly; "has he not been sufficiently punished?"

"Punished!" said Bernhardi, fiercely; "do you know what he attempted?"

I shook my head.

"Then it shall for ever remain unknown. Punished! —one short minute, and I should have been too late! Hear you that? Will you act for me? Will you act now? Will you see that we meet forthwith?"

"That will depend on your adversary."

"Oh, I cannot wait—I will not wait!" exclaimed Bernhardi: "go! go!"

The irresistible frenzy of the student prevailed. I was taken by surprise. Quiet and peaceful as was the life I led, before I was aware of it, I found this strange commission thrust upon me; and almost before I knew it, I was in Balaiguer's room.

The Marseillese sat smoking, with a light cap upon his head, which only partly concealed some recent bruises.

"So," said the savage, "you come to have your laugh with the rest! and *you* were the tell-tale, eh?—you were the sneak!"

"We will settle these epithets by and by; at present another's business has a preference. You must be aware that your conduct this morning——"

"What of it?"

"Nothing, except that Bernhardi will meet you at

any moment you will appoint; for him the sooner the better."

"For me the sooner the better," growled the Marseillese.

"Who is your friend?"

"*Sacré bleu!* that remains to be seen. I will send him to you."

I went back to my room, somewhat surprised at the bold bearing of Balaiguer, for I was sure that he was a coward, until I remembered that he was an expert swordsman, and that Bernhardi once told me that he himself had little knowledge of the weapon.

In about a quarter of an hour an acquaintance called on the part of Balaiguer. As I anticipated, swords were chosen. As to time and place, the Marseillese was quite indifferent.

There was a large hall over a billiard-room in a street near by, where many of the students were in the habit of fencing, but where, at that hour of the day, no one was likely to be seen. To this hall we agreed to repair forthwith.

I summoned Bernhardi, and, accompanied by another friend, according to arrangement, we proceeded to the appointed place.

The German grew more and more excited. Never had I witnessed such an awful manifestation of human passion.

"Are you expert with the small-sword?" said I, as we went along.

"It matters not how *expert* I am; I shall pass my weapon through his heart!"

These words were spoken slowly and deliberately, yet the speaker was boiling with rage.

We entered the hall. Balaiguer and his friends were on the spot. Bernhardi took no notice of any thing. His eyes glared more horribly than ever; a white foam gathered on his lip.

Balaiguer seemed in spirits. He was evidently delighted at the excitement of his adversary, and confident in his own skill.

The preliminaries were soon settled, (for a student's duel was no very serious affair, it rarely being a matter of life and death, generally ending in a scratch, or at most a flesh-wound,) and the parties stepped forward for the encounter.

I looked at Bernhardi with a curious eye. His "case" was a phenomenon in physiology; for excited—nay, almost raving—as he was, I perceived that physically his muscles were firm; there was no tremor in a single nerve. Dupuytren himself, at the moment of commencing the most serious operation, never carried a firmer hand. When he looked his adversary for the first time in the eye, he could scarcely contain himself.

The signal was given.

"Beast!" screamed Bernhardi, as he brought his sword awkwardly to a guard, "shall I kill you at once, or shall I do it with a 'one, two, and three?' Is a moment's time worth any thing to you? If so, you shall have it; for a moment saved *her!*"

Balaiguer smiled triumphantly at this new proof of his adversary's frenzied state, and made an ordinary pass with which to commence the combat. Their swords met for the first time.

"Now for it!" said Bernhardi. "One," (a pass, parried by Balaiguer;) "two," (parried also;) "three!" The Marseillese fell, thrust through and through!

Bernhardi gazed at the dead man for an instant. "Dog!" he exclaimed; then, throwing down his sword, he clutched my arm, and clinging to it convulsively, he tottered down into the street.

I supported him to my apartments. He was as weak and powerless as an infant. In the course of an hour he regained sufficient strength to walk home without assistance, and extorting a promise from me to visit him the next morning, he went away.

I bolted my door, and, throwing myself into a chair, remained the rest of the afternoon and all the evening sitting quite alone. At length I went to bed, but I could not sleep. Whichever way I turned, the form of the Marseillese, cold, stiff and stark, lay stretched out before me. The fierce whiskers, the grim moustaches, and the savage beard, curled as fierce and as grim and as savage as ever, as it were in mockery of the pallid features they once so gayly adorned; while close at hand stood Bernhardi, his sword dripping with blood, the very incarnation of an exulting fiend. Not for one minute did I close my eyes the whole night, for when I attempted it the images grew more horrible, and I was forced to open them in order to dispel the illusion.

I tried to believe the whole a dream, that I had been oppressed by a horrible nightmare. I could not realize that I had been so suddenly arrested, turned from my quiet, unobtrusive way of life, and made to participate in the death, not to say murder, of a fellow-creature: it seemed as if the morning would bring some relief, and for the morning I anxiously watched.

It came at last, but I was in no haste to stir out. At length a knock at my door roused me. It was the young Frenchman, and I rose to admit him. He told me about what I feared to ask. Balaiguer was discovered early in the evening by some students who repaired to the hall to fence. They gave the alarm, and the police took the matter in charge. Three students, acquaintances of the deceased, were missing; (they were the two friends of Balaiguer and the young man who with me acted as friend to Bernhardi, who fearing the annoyance, if not the danger, of a legal investigation, had immediately left Paris;) it was understood that Balaiguer must have fallen in a duel, and it was a natural conclusion that the three who fled were his antagonist and the second of each party. So suddenly had the affair sprung up, so suddenly had it terminated, that not a soul beyond the persons present, except the young Frenchman, who could guess the truth, knew or suspected any thing relating to it. The latter now begged me to rise, and appear as if nothing had happened, and insisted that I should take my coffee with him.

I asked for Bernhardi. The young Frenchman had not seen him, but, singular to say, his name had not been mentioned in connection with the tragical affair. Two strong cups of the best coffee, with the usual accompaniments of a roll, two eggs, and a plate of fruit, did much to restore the steadiness of my nerves, which had been, I admit, considerably shaken.

Recollecting my promise to visit Bernhardi, I crossed over soon after breakfast to see him.

He was standing at the door of the *conciergerie*, apparently waiting for me.

He took my hand as I came up, and inquired anxiously how I was. As for himself, his countenance had resumed its pale, saddened expression; no trace of the passions, which had been so terribly roused, appearing there.

He requested me to go with him to his room, and I willingly assented. We entered it in silence. Bernhardi pointed to a chair, and I sat down, while he took a seat near me. I glanced over the apartment. It bore traces, all around, of the presence of — woman. It was furnished with admirable taste, and ornamented with pictures, engravings, and embroidery. Folding doors, which however were closed, led into another room, and with the one we were in, evidently formed a suite. I had scarcely time to finish this rapid inspection, when one of these doors opened, and, I speak considerately, the loveliest, most angelic-looking being I ever beheld, entered. Her face was as faultless as the Madonna of Correggio, her form as perfect as the Venus of Phidias, her countenance absolutely lovely and serene; her eyes were a deep hazel, and the heavy tresses of her rich brown hair were exquisitely braided over her temples, and wreathed around the back of her head. She walked slowly forward, and, as if unconscious of my presence, approached Bernhardi, and throwing her arms over his shoulders, pressed him fondly, while she exclaimed, "Dear, dear Ernest, have you returned at last? Oh! do not go out again!"

Bernhardi shrunk from the embrace as if suddenly

bruised by a blow, while his countenance exhibited signs of physical pain and suffering. He rose quietly from his seat, and, putting his arm around the lovely intruder, led her gently back to her apartment, without any resistance on her part. As she was leaving the room, she turned her eyes casually upon me; at once a horrible suspicion darted through my brain, my heart beat violently, my knees shook together, I almost gasped for breath. Bernhardi closed the door and resumed his seat by me: his countenance was troubled; he looked in my face sadly; after a while he spoke.

"I asked you to come here that I might give you the explanation to which you are entitled. Rumour and gossip have doubtless been busy with me. I care for neither, and although I have no desire for notoriety, I am indifferent to it. You have laid me under an obligation which I can never remove, and one which peremptorily demands that I should explain all to you. I shall be brief, just as brief as the bare recital will permit. Will you listen?"

I bowed assent.

"I am a native of Wirtemberg. I was born in the little village of ———. My father was a wealthy peasant, and I am an only child. I was brought up tenderly, and as I was said to manifest considerable wit and intelligence, my father determined to educate me. In the same village dwelt a widow lady, whose husband had been an officer of some distinction under Napoleon. Upon his death his widow had come back to her native place, bringing with her an only child, a little daughter of some seven or eight years of age. I was then about ten. The widow's fortune was small, but sufficient for the simple habits of the place she had chosen for her home. My father had known her when a young girl, and with my mother often called at her little cottage. In this way Rosalie and I were thrown much together. Indeed, after a while we were almost inseparable. In our sports and plays I was always Rosalie's bachelor. I used to call Rosalie my little 'wife' and she called me her little 'man.' This was without any reflection on our part: neither of us were old enough to think seriously.

"At length the time arrived when I was to go away to school. I suppose I was twelve years old, and took leave of Rosalie with a heavy heart. I really think at that early age I *loved* her. Well: years ran along. From school I went to Heidelberg. I was ambitious, I was full of energy, and my love for Rosalie preserved my boyish purity of heart. Year after year, as I visited my home, I was surprised to find in her some new grace, some new charm, some new beauty. At sixteen, she seemed to me all that could be imagined of what is lovely and beautiful. A delicious ecstasy floated through me when I felt that she would one day be mine.

"But I had a drawback to my happiness. In spite of every effort to believe the contrary, I could not feel in my very heart that I was loved by Rosaile even as I loved. True, she was fond of me, but it seemed rather the attachment to be felt for a protector or a brother, than the devotion of love to love.

"I nursed myself with hopes. *I* had never loved but Rosalie; no one had ever loved me but Rosalie; and who could expect that a young girl should show the same deep devotion that marks a powerful, manly heart? This was the way I reasoned. Rosalie, I was certain, kept nothing from me. She told me every thing. She said she loved me as well as she loved her mother; ought I not to be satisfied? But when I pressed her to my heart, I felt not that electrical affinity which cements in *one* hearts which *are* united; still I did not complain: how could I complain, when Rosalie told me I was all to her?

"I passed three years at Heidelberg, and then went to Munich. having determined on medicine, I prepared to follow the study with devotion. I had been at Munich nearly a year, and I yearned to come home and see Rosalie. I had stayed away longer than usual, because I wished to take a degree in my profession;

then I felt that I could claim Rosalie for my wife. I did go home. Let me hasten my tale. I greeted my parents; every thing was well. I hurried to Rosalie; she was well too. She ran out to meet me. She was delighted to see me. Never had she looked so beautiful. As we entered her mother's house together, she exclaimed: 'We have a guest—a charming guest; a son of my father's dearest friend. He has been with us for a month, but must soon return to Paris; and I shall miss him so!'

"My brow grew overcast; my heart sunk. I said nothing; I believed my destiny sealed. I did not even look upon Rosalie reproachfully. How could I look reproachfully upon *her?*—for her soul was pure; it knew no guile; it was incapable of concealment, or coquetry, or caprice.

"Suffice it to say—for the narration is too much for me—that on entering the cottage I found a young and handsome French officer. He was, as Rosalie had said, the only child of her father's dearest friend, and had sought out the widow at his father's request. Hear me," whispered Bernhardi, while he drew his chair nearer to me. "I made friends with that young officer. With the closest observation I sifted him as wheat: I found him honourable, high-minded, good-tempered, pure. I satisfied myself that Rosalie loved him, (poor child!

she did not know it;) I sought an interview with Ernest de Fleury—that was his name; I pressed the secret from him, which he swore should otherwise never have been revealed, for he knew that Rosalie was my betrothed. Then I turned, and went for Rosalie. I had a long, long interview with her. For Heaven's sake, let me hasten!" gasped Bernhardi. "You—you—guess the rest; guess it *all*. The sweet angel was sweeter than ever; but—but—I got at the truth. She protested that she would never, never give me up; those were the words, '*give me up*.' That was noble; and then she pitied me; but I was not to be thwarted. I took her with me to the cottage. Ernest de Fleury was there. I joined their hands and ran out—I ran home, and—and—old as I was, I threw myself into my mother's arms, and burst into tears. Oh! GREAT GOD of this strange universe! what is like unto a mother's love? There I sat all of the day—all of the evening—my head pressed against the breast that had given me life and nourishment, and there, in broken sentences, amidst sobs and tears and groans, I told her all. And my mother, how she sympathized with every heart-pang! how entirely did she understand my feelings and my motives! how tenderly did she intwine her arms around me, until at last I fell asleep upon her bosom!

"The next day I returned to Munich.

"How long I should have remained away I know not; but at the end of a twelvemonth I heard from my parents that a fearful epidemic was raging in my native village, and that they desired to see me. I went home. The village was in mourning; a malignant fever was carrying off the inhabitants. Rosalie's mother had just expired, and Rosalie herself lay sick unto death. My parents had thus far escaped.

"I went at once to Rosalie's cottage. I became her physician, attendant, nurse. I watched night and day. The fever had reached its height, the crisis had come, and Rosalie opened her eyes on the fearful morning which should decide her fate. I saw that she was saved. A grateful look of recognition beamed in her countenance. She was very weak, but the danger had passed.

"The next morning fatal news came to the village. A letter to Rosalie's mother, now no more, announced the death of Ernest de Fleury. He had been seized with '*la grippe*,' then the prevailing epidemic in Paris, and had died in six hours.

"Rosalie was the first to see the letter. One glance was enough; she fell back in my arms, in violent convulsions.

"Days and weeks and months I watched by her bedside. At length her strength returned; the bloom once more freshened her cheek. I was full of hope. One morning, as I entered, she sprang up from the bed, and throwing her arms around me, she exclaimed, (as you heard her exclaim but just now,) 'Dear, dear Ernest! have you returned at last? Oh! do not go out again!'

"Then my cup of misery was full. My Rosalie, Ernest's Rosalie, was—*imbecile!*"

Bernhardi paused; he spoke not a word for five minutes; then he said: "You know the whole. She thinks that I am her Ernest, and she is happy in my presence. Physically, she enjoys the extreme of health; mentally, alas! she is no more! I came with her to Paris, hoping that the change would benefit her, for Ernest lived here; but it is of no use. My prayer is that my life may be spared to outlast hers; for what will become of her when I am no more? Do you blame me for assuming the execution of the law upon that wretch? You cannot blame me. I blame not myself.

"My life is devoted to her. I honour my MAKER, who has given in CHRIST JESUS the great example of a disinterested love. Who is so selfish as to whisper to me that 'love must be mutual?' I acknowledge the devotion of woman. I know that often she dies of a broken heart; but I *live* broken-hearted!"

Bernhardi had finished. I took his hand and pressed

it in silence, and came away. That afternoon I quitted Paris *en route* for Italy.

On my return here, after the lapse of more than a year, I made inquiry for Bernhardi, and learned that, several months before, he had left the city with the unfortunate Rosalie, and had gone no one knew whither.

CHAPTER IV.

RAMBLES OVER PARIS.

EVERY day I took a walk by myself over some portion of the city. My plan was desultory, but not irregular. There was no method, yet there was a purpose in it, viz., to know what was going on in Paris. Perhaps strolling about the streets was not the best way to find out, but none better occurred to me.

In these walks I was continually stumbling on objects of interest, or chancing on some little adventure. I would sometimes drop quietly into the little shop of the *charbon* and faggot-vender, and listen to the history of his trials and struggles—for all *charbon* and faggot-venders, be it known, have their trials and struggles;—besides, I was interested to learn the cause of the extravagant price of fuel, for in cold weather our pockets were nearly drained in the attempt to make ourselves comfortable.

I frequently introduced myself into the little niche, where a smiling, cheerful, and vivacious cobbler hammered away from morning to night under the protection of the Holy Virgin, whose image, adorned daily with fresh garlands, was placed directly over the entrance.

A rare fellow was Jacques Tourneau, with whom the world always wagged happily and well: with a pleasant word for every body, a joke for all occasions, and keen perceptions to season it, with a good-tempered wife, (taking his word for it,) and half-a-dozen healthy children, Jacques was, all things considered, the happiest fellow I met in Paris. I learned many philosophical lessons from Jacques Tourneau.

Occasionally I would stroll into a church and see what sort of persons in Paris were devout. Then, perhaps, I would take a walk in the gardens of the Luxembourg, for I confined myself, generally, to our own side of the river. I confess that the gardens attracted me greatly. A great variety one could see there. Old, and middle-aged, and young, with scores of children, sitting, walking, running, frolicking. A rare place for me were these gardens of the Luxembourg!

Then, again, the *Hotel des Invalides:* I have passed hours quietly watching the veterans who lounge about the grounds. Especially can I now call to mind a sturdy old fellow, with *two* wooden legs, and but one arm, and marks of many a cut upon his face, who used to sit just two hours every day on one of the stone benches, in the *Place de Vauban,* fronting the Hotel, and who, with countenance calm and unmoved, placidly contemplated whatever passed around him. I never saw him exchange a word with a brother soldier, although frequently seated on the same bench. Had I not feared some misinterpretation of my motives, I would have addressed him. But there was something in those truncated limbs, and in that scarred visage, which forbade an ordinary intrusion.

The inmates of the *Hotel des Invalides* wear no appearance of disappointment or discontent. They feel that it is an honour to be pensioners of *France;* so that one beholds no forlorn looks, no depressed glances, nothing, in short, of that unpleasant expression of countenance which is almost always observed in retreats for the decrepit and the old.

The stranger who visits the chapel of the *Invalides* will encounter few of the inmates, unless at the time of service; but there are always a small number who can be seen kneeling, repeating a prayer, or going through with their *Ave, Credo,* or *Confiteor.* After a "fitful fever" of marches and assaults, of sieges, sorties, and pitched fields, of fierce pursuits and sullen retreats, of bloody defeats and bloodier victories, it *is* a touching sight to behold the soldier kneeling before the cross, asking forgiveness and absolution.

I observed an elderly officer, who appeared much

superior to the majority of his *confrères*, and who came very regularly to the chapel. He was about fifty, tall and slender, with a serious countenance, and an air of habitual depression. He used to kneel with so much devoutness, and repeat the prayers so earnestly, and afterwards come away with a look so melancholy, that it touched me to the heart to witness it. He had not been wounded, so far as I could see; he had lost none of his limbs, but his face was pale and wasted, and loose, straggling gray hairs were scattered over his forehead.

How much it adds to the intenseness with which we regard misfortune or calamity, to separate some individual object, and fix our attention on it! I believe one could easily become utterly miserable by this very process. I have myself, in this way, on many occasions, been made wretched enough, and only escaped by turning to the brighter scenes of life. So it is always; light and shade—light and shade again. But without light and shadow can there be a *picture?* There is, at the same time, a fascination in the contemplation of great suffering difficult to explain. Perhaps it may be traced to the unconscious sympathy we feel with whatever is intense, whether it be ecstatic or agonizing, and which underlies almost every other emotion. . . . .

On one occasion, in turning to leave the chapel, when

I was standing near the door, the melancholy officer of whom I have spoken dropped his handkerchief. I picked it up, and observed, as I took it in my hand, that it was of a description used only by ladies. I stepped at once towards the owner, and gently touching his arm, I said:

"Your handkerchief, sir."

A faint, hectic blush overspread his cheeks.

He seized it almost eagerly, gazed at it an instant with much tenderness, as though it were some dear object, and put it in his bosom; then taking my hand in both of his he pressed it silently.

"I am very glad," said I, "that I discovered it in time."

"It was my wife's."

His lip quivered slightly, but he showed no other signs of emotion. Still he retained my hand.

"Forgive me," I exclaimed, "I have intruded on feelings which are sacred."

"Monsieur shows that he has a heart."

He pressed my hand once more, bowed low, and walked away.

I do not think I can ever forget that old French officer. Although I used frequently to see him after this occurrence, I never accosted him again. Yet I busied myself, at times, imagining what had been his peculiar griefs.

His *wife*. It was his *wife's* handkerchief. Her memory was all he had to cling to. Children none: relatives none. *She* had been to him his sole and only friend, and she was gone. That was it. Perhaps—I carried my conjectures further—perhaps *he* had not been as affectionate, as constant, as kind, while she lived, as he now felt he ought to have been, and, like too many who do not

"———understand a treasure's worth
Till time has stolen away the slighted good,"

he had appreciated her *too late*. Perhaps he was now tortured by a recollection of her last sad, yet not reproachful look, and cherished, as a part of his existence, a tender though unavailing remorse. But whatever might be his personal history, I felt an assurance that his daily prayers and supplications were not put up in vain.

I have mentioned the gardens. The most joyous sight to be met with in Paris, is that of the children who congregate there; hopping, running, skipping, playing puss-puss-in-the-corner, (a tree for each corner,) and even blind-man's-buff. As my friend Clements remarked, it always seems as if French children were very precocious to have acquired a *foreign* language so young.

There was one charming, ruddy, brown-haired little

creature, about four years of age, who interested me greatly. She was so full of childish spirits; her laugh was so clear and so mirthful; her voice, though infantile, was so sweet, and her motions so light and airy, as she flew from spot to spot, that I became absolutely fascinated. An elderly woman, plainly dressed in black, sat always on one of the benches near by, engaged usually with her needle, or in knitting. I observed that she watched the child's movements continually, with eyes beaming with affection. Could she be the mother? Certainly not. The nurse, perhaps? No. I was not satisfied to call her the nurse. She did not wear the expression which smacks of service, and which is generally unmistakable.

I seated myself one day on the same bench with the good dame. "What a beautiful little child!" was my first observation to her.

"Which one, *Monsieur?*"—She knew very well, without asking.

I pointed out my favourite, who, with several of her playmates, was frolicking a few steps from us.

"Ah, that is my little Annie, my grand-daughter."

"Indeed! and its mother?"

"She is all I have left, *Monsieur*."

The French have more delicacy than any other people in conveying a melancholy idea.

"How you must love the little creature!" I exclaimed, involuntarily.

"Indeed, *Monsieur*," she replied, "I see my lost Annie living her life over again; she is the very same, just as she looked, just as she acted."

At this instant little Annie ran up, and bounding into the old lady's lap, cried, "Mamma, I have something to tell you—hold down your face;" with that she gave the ear, which was thus brought within reach, a sly pinch, slid down, and darted away; she returned almost in the same moment, resumed her place, kissed the "poor little ear," as she called it, and once more ran off.

"Just as I was saying to you, sir, she has all her mother's sweet ways, and I have taught her to call me 'mamma,' and it seems—— but no, I cannot lose sight of my *child*, my first Annie, who *was* like this one, and who grew up to be a girl, and then to be a woman——."

The old lady's eyes filled with tears.

"And she died?"

"Her husband died first. That nearly killed her. Then she took a fever. I did all I could—nothing availed. I nursed her—I gave her every thing with my own hands, and she would say, 'My mother, do not do this, you will fatigue yourself; I feel easier now; go—do go, and get some rest.' But I could not leave her. Sometimes she made me recline on her bed, and put my arms around her, and then she would look into my face and smile. Oh, could you but have seen that smile! . . . . Alas! nothing could save her. We had a noted physician from the *Hotel Dieu:* he would come two or three times a day, and take hold of Annie's hand and say, 'My poor child, what makes you so sick?' Then he would encourage her, and speak so kindly that I could have fallen on my knees and blessed him. He was with us when she died; he wept like a child, and——"

The recital was too much for the poor woman. She placed both hands before her face, vainly endeavouring to prevent the tears, when little Annie, happening to see it, ran towards her, all in a glow as she was, and, springing into her favourite place, threw her arms around her grandmother's neck, and by every term of endearment and affection, by kisses and caresses, attempted to moderate her grief.

It was more than I could endure. I turned and walked hastily from the spot; *my* eyes were moist too, and once away from observation, I drew my handkerchief from my pocket and wiped them.

CHAPTER V.

STUDENTS' NONSENSE.

A CLEVER knot of young fellows were assembled around the door which led into the garden adjoining the house in the *rue Copeau*. I do not know why students are so much in the habit of congregating around the threshold of an outer door. Such is the fact undeniably. Who will undertake to explain it?

It was a fine, pleasant day, in the fall of the year. The leaves were beginning to drop off, and the air was autumnal. One by one, as they left the *salle-à-manger*, the young men passed out into the garden with pipes, meerschaums, and segars; some with books in their hands: most wore caps, but a hat here and there could be seen on the head of some resolute American, who in this way showed his contempt for prevailing customs.

Of the company, one was a Pole, two were English, three American, two German; there were also an Italian, an Irishman, and a Genoese, besides several the place of whose nativity had never transpired. They were, for the most part, diligent students, somewhat reckless of the ordinary demands of society, but having a decided purpose in view. The majority were studying medicine.

The Irishman was a Roman Catholic, and devoted himself to theology. His name was James Daloney. Where he now is, I do not know. He was about taking orders, and is, doubtless, labouring some where in his holy calling. Should his eye chance to fall upon this page, I beg to send him a friendly greeting, for I am sure he will not have forgotten his sojourn in the *rue Copeau*, nor his companions there.

One of the Germans was named Franz von Herberg. He was a painter, devoted soul and body to his art. He was open-hearted and sincere, somewhat sensitive to criticism, refined in character, of an exquisite humour, yet subject to frequent depression of spirits.

The other German, Jacob Wahlen, was a student of philosophy, full of mysticism and Spinoza.

The Italian and Genoese—so they were always named —came to the house together, and were much in each other's society. They had incurred, I imagine, in some way, the resentment of their respective governments, and were now exiled.

The two Englishmen were as unlike each other as was possible for two persons to be. One was conceited, and a cockney; the other was my delightful friend, Clements.

Vincent, Partridge, and myself, with three or four others, completed the group.

"What is the news to-day?" said Vincent. "Has any one been on the other side? is Louis Philippe recovering?"

No one knew.

"I was down in the country yesterday," said the cockney. "Lord Roslin, the brother-in-law of the cousin of our ambassador, invited me. 'Pon my word, we had such a capital time. I am to go out shooting with him next month—such a box as he's got: he's such a sportsman, too; he told me he shot thirty-three hares in England one morning before breakfast."

"He must have been firing at a wig," said Partridge.

A general laugh followed this sally, which the other did not seem to comprehend, for he went on in the same tone, not heeding the interruption.

"By the way, Franz, when are we to see the new painting?" asked I.

"Never, I fear," said Franz; "I have tried to paint the man, and——"

"You can't get the right *expression*, I suppose," said Daloney.

"Go to the *Morgue*," said one.

"Or to the public executioner."

"You should have been here in '30," said the Italian;

"that would have been a time for taking dead men in all shapes."

"Gentlemen, you don't understand me. You speak as if I wanted to get upon my canvass the characteristics of *death*; that, I admit, I can find where you suggest: but it is the *living* expression which sometimes lingers on the face *after* death that I would transfer. Bah! 'tis not so easy to put the two things together."

"That's not the only disappointment which Franz has met with lately, in putting two things together," said Daloney.

"Ah! how is that?" cried several.

"Why, our friend here undertook to paint a cow and a cabbage on the same canvass, and both were so natural that he had to separate them."

"Bravo! bravo! Daloney;" and there was a general shout.

"Daloney," said Vincent, gravely, "take my hat. I never will wear one again."

"It comes in good time," whispered Clements, loud enough to be heard by the whole party, while Daloney gave him a glance to be silent.

"No, no; it is too good to be lost," said the other. "You must know, gentlemen, that yesterday our friend treated himself to a new hat; price, nine francs, fifteen sous, and two centimes. Instead of coming home, like a rational creature, to his dinner, he wanders into the *rue Rivoli*, dines, takes *café*, and rises to depart. His hat is missing; he looks about quietly; he is sure he placed it on the seat just behind him; he looks again; he discerns a dirty piece of paper with two lines scrawled on it; he picks it up and reads as follows:

"'I have taken your new hat—but I leave you my eternal gratitude.'"

Another general laugh succeeded Clements's narration.

"You have interpolated," said Daloney; "there was not one word about gratitude, else I had been satisfied; there was nothing, in short, for my fine beaver, but an old shabby, torn specimen of a *chapeau*, not fit for the beasts of the field to wear."

"They *would* look well in hats, to be sure," said Vincent; "don't you think so, professor?" turning to Wahlen.

"I don't think, so soon after dinner. It disturbs my digestion."

"How solemn you grow! Pray, Franz, let's have the story about Wahlen's going to see the juggler."

"Ja—ja—you may tell it in welcome," said Wahlen, seriously, "if it will pleasure the company."

"Oh, *do* let's have it, Franz," cried half-a-dozen.

"I can give it in word. Wahlen and I went to see a juggler who exhibited on the corner near the Odeon. We had front seats. In the course of the performance he asks some person to step on the stage to assist in a piece of *diablerie*. He beckons Wahlen, who at that moment was thinking of any thing but what was going on. Wahlen starts at once. Among other things, he asks Wahlen to hand him a napoleon. 'You see,' cries the juggler, addressing the audience, 'this gentleman hands me a napoleon. I put it in my pocket. Now let every one watch me narrowly. *Siberah, Vibberah, Tintentuncleristhatch—Presto, Voila!* The gentleman will tell you it is in his pocket again,' appealing to Wahlen, who was at that moment deep in Fichte, or Jacob Boehme, and was startled into saying, 'Yes,' before he knew he had said any thing. The juggler, with most triumphant air, now moved our friend to take his seat."

"'Please return me my napoleon,'" said Wahlen.

"'Swindler!' exclaimed the juggler, in a low but resolute tone, 'have you not *said* publicly that you had it back again? If you make the slightest disturbance, I will have you turned out of the house.'"

"And I *made* no disturbance," interrupted Wahlen, "for two reasons. First, I was properly punished for forgetting where I was, and what I was doing; and, secondly, the juggler's unparalleled audacity deserved its reward."

"Ah! Jacob Wahlen," said Vincent, pleasantly, "you

are a perfect mystery. You will become in due time a great German professor, and when you die—distant be the day—you will doubtless say, as your admired Hegel said, 'I shall leave behind me but one man who understands my doctrines, and he does *not* understand them.'"

"Perhaps," ejaculated Jacob Wahlen; and having uttered this single word in reply, he was again deep in his philosophical revery.

Here three or four of the company went across to the billiard-room.

"Well, Franz, are we not to see the picture after all?" said the Italian.

"I tell you the truth, Signor Italiano, I cannot paint it. I have sketched and rubbed out, and sketched again —it's of no use."

"Why don't you do what some of your craft have done before you?"

"What is that?"

"Drive a trifling bargain with the old gentleman down stairs."

"I won't do that. I believe in the devil, but don't think him a good artist—he colours too highly."

"You must admit he *draws* well," said Vincent.

"He's not the subject for a joke, at any rate," replied Franz.

"Franz is low-spirited, I do believe."

"Supposing he is," said Clements, " it is as it should be. You know the saying—'Melancholy is the characteristic of the German — wit of the Frenchman — gallantry of the Spaniard—love of the Italian—and, I am almost too modest to add—sense of the Englishman."

"While a happy combination of all, you find only in the American—ahem," said Vincent, laughing. "But come, Franz, permit us to run up into your rooms and see what you have done."

"You shall, with pleasure, but the picture I cannot show you."

Three or four of us accordingly followed our friend to the top of the house, where, of course, we had been often before. The appearance of the room was like that of every artist. One beheld the usual arrangement for light, the easel, stands for paints, &c., one or two unfinished pictures about the room, a few exquisite old paintings, and several pieces placed on the floor and turned to the wall.

"Now, won't you change your determination and show us the picture, although it be unfinished?" said Vincent.

As he said this, he took hold of one of the larger pieces of canvass which was placed to face the wall, and, I imagine quite involuntarily, turned it around.

An exclamation of horror fell from every one, succeeded by a breathless silence as our eyes were *fixed* as if by enchantment on the painting.

It was that of a young girl, no more than seventeen,—having a classical face, with dark hair and eyes. In saying this I have said nothing. It was the expression which made the painting what it was; and yet there *was* no expression which one should recognise as human: and as for the eyes, they seemed, while you looked at them, *to creep into you.*

While we were thus standing transfixed, Franz rushed forward, and seizing the picture turned it back again, exclaiming, "For Heaven's sake, not that—not that!"

"Ah, my dear fellow, you are not yourself this evening; we will not tease you any more,—but pray tell us what moves you so?" I said.

"The fact is, the black dog has been sitting all day on my left shoulder, as my Scotch friend Macdonald used to say. I do not know why or wherefore; and now you have turned around that picture, which has not been touched for a twelvemonth, I shall carry two black dogs instead of one—perhaps it will help to balance the load. At any rate, I will show you the unfinished thing you came to see, although I said I wouldn't. It will create a diversion at least."

"No, Franz," said Clements, "you did not wish us to see it, and we will not look at it. But we have a request to make—I think I can speak for the rest. We want to know if the picture we have just seen is drawn from life?"

"I perceive," replied Franz, in a more cheerful tone, "that there is no escape for me. Whoever sees that picture once, never rests till every thing is told. For this reason, I always keep it with the face to the wall, and usually with something thrown over it; and, as I told you, I have not seen it before for a twelvemonth."

"How could you ever have painted it?"

"*Me?*" replied the artist, with a look of terror. "Mother of Heaven! I did not paint it! No, not I." And Franz von Herberg stared at us for a moment as if he had forgotten who we were. He quickly recovered, and said, hastily, "Sit down—sit down; you shall hear what I have to tell about that painting. But, in the first place, let me ask if any one of you wishes to examine it more closely; if so, you are to do it before I commence, for when I have finished you must not ask to see it."

No one expressed the least desire for another look: so fearful, I may say so terrible, was the effect of the first sight upon each one of us. Whereupon Franz took the picture, and, without changing the position, placed it in his closet, and threw a quantity of loose papers over the canvass. Then bolting the door, he drew his chair towards us, and commenced as follows:—

CHAPTER VI.

THE TERRIBLE PICTURE.

"LIFE is not a particular form of body, but the body is a particular form of life. The body relates to the soul as the word to the thought." So says old Jacobi. He did not address artists, but artists may learn a lesson from the saying. So may you, *Messieurs* students of medicine. For myself, I always carry it in my head.

I don't know why I commence by quoting Friedrich Jacobi, when I am to tell you about Ernst von Wolzogen, except that it was a favourite saying of Ernst, and since— but no matter.

Ernst and myself were born in the same village. He was but a year older than I, and we were placed at the same school together. From his childhood, Ernst manifested a strong love for his art. At that period I had but little idea of it, and I owe to my intimacy with him my taste for painting. With a handsome person, eyes black and piercing, with long, dark hair, and a magnificent brow, he certainly was the handsomest fellow I ever saw. As an artist, he was bold, independent, full of original conception, no imitator, no copyist, no follower of any school, although he appreciated, as much as any one, the works of the great Masters, as they are called. From the first, he was remarkable for throwing *the very living thing itself* upon the canvass, in a manner which would astonish us all. There might be errors—there were errors, of one kind and another,—but, for all that, the thing itself stood before you. It mattered little whether it was a portrait, or a landscape, or a historical piece; the effect was produced. When certain faults were pointed out to him, he would say, "I know it—I perceive it—I will mend it by and by; but first I must see that my picture is *alive*, that it is *real.* 'Life is not a particular form of body,' &c.; the rest will come soon enough. We must have patience. It *will* come."

Away from his easel, Ernst von Wolzogen was dreamy and superstitious. He was susceptible, too, but very shy, so that before he was one-and-twenty he had fallen in love and had his heart broken a dozen times without so much as speaking to his *inamoratas*. Once at his labours, however, all the unhappy mists which gathered about his brain were dispelled; then, and then only, he was really himself.

"ART, my dear Franz," he would exclaim, "Art belongs to man only. In Art there is no divided empire:" and he would triumphantly recite those lines of Schiller:

> In diligent toil thy master is the bee:
> In craft mechanical, the worm that creeps
> Through earth its dexterous way, may tutor thee;
> In knowledge, (could'st thou fathom all its deeps,)
> All to the Seraph are already known:
> But thine, O MAN, is ART—thine wholly and alone!

I have said he was superstitious. I can hardly expect to be credited if I tell you what a slave he became to all sorts of signs and omens and prognostications. He believed, too, in presentiments and warnings. He credited ghost stories and tales of apparitions, and maintained that, were it not for our gross organization, we should all enjoy the privilege of second sight, and I do not know what else. This had a very unhappy effect on him—an effect I was quite unable to counteract, although we were bosom companions and had been almost inseparable from the time we commenced our studies.

"My friends," continued the artist passionately, after a moment's pause, "I loved Ernst. I loved him for these very weaknesses, which betokened a spirit far removed from this earth. Beyond every thing, I loved him for his appreciation of our artist-life, and for having roused my soul to a proper sense of it."

As I had much more of the practical in my composition than my friend, it fell to me to look after the economy of our every-day life, while he endeavoured to carry me along with him in the rapid strides he was making in his art. We went over Europe in company. We dwelt together in Rome, in Florence, in Naples, in Vienna, in Munich, in Dresden, in Paris. We accompanied each other to see paintings and statues, and, in short, every thing worthy of examination.

We had spent some time at Dresden, and Ernst was becoming more and more subject to the unfortunate influences I have named. I proposed, therefore, as an agreeable change, that we should go to Paris, and take apartments in a pleasant part of the town, and thus try the effect of gay and lively scenes. There was at the same time a painting in the Louvre—a landscape by Annibal Carracci, which had lately been transferred to that palace, which we both wanted to see.

We came to Paris, and took rooms in the *rue de la Paix*. The first morning after our arrival, Ernst started out alone to take a stroll through the gallery of the Louvre, in order, as he said, to report about the "landscape." He promised to return in an hour or two; but he did not come back till quite late in the afternoon. He was in a state of most cheerful excitement. He had not looked at the "landscape," but he had seen the most exquisite of all living pictures.

Ernst was always extravagant when describing his favourites, but he now exceeded any thing he ever before said in praise of female perfection.

"Her name?"

He did not know—he did not want to know. He only wanted to gaze on her, to be inspired by her, to worship her.

"I suppose," I said, "I may be permitted to visit the gallery and steal a single glance at the fair one."

"Indeed," replied Franz, "you *must* see her; otherwise you have a right to think me beside myself."

The next day we went to the gallery together. We passed nearly half way through the hall when Ernst touched my arm.

Seated before the painting by Teniers, of the "Village Wedding," was a young girl, scarcely more than seventeen. Her hat and shawl and gloves were laid aside, and she herself was so completely absorbed in transferring the scene to her canvass, that she did not appear aware of any thing that was going on around her.

She was indeed a beautiful creature—perfect, it would seem, in form and feature, and apparently of great simplicity of character; and no one could witness the enthusiasm with which she pursued her employment without feeling a strong interest in her. A man-servant, in plain livery, stood behind her. This indicated the enjoyment of competent means, while a certain indescribable bearing evidenced that our young *artiste* was of gentle birth and breeding.

"What shall I do?" whispered Ernst. "I must turn copyist. Let us see; what is the next painting? 'The interior of a smoking tavern.' Pshaw, that will never do; but on the other side? Ah! 'Diogenes with his lantern looking for an honest man'—Rubens. I'll copy it. By Jove, I'll copy it. But is it honourable to take such an opportunity to be near this charming creature? is it a fair advantage, think you?"

"Why not?" I replied; "surely, we may admire all the portraits here, whether on canvass or not; and you have certainly a right to select your position."

I wish you could have seen the work Ernst made of copying the piece he sat down to. Sometimes his Diogenes stood out with long, black tresses, and a delicate lithe form: again the cynic would absolutely forget his lantern, and at another time omit to light it. Droll business was it for Ernst von Wolzogen, already the pride of the younger German artists, and the admiration of all who saw his productions.

The young girl, meanwhile, was busily engaged. Acute as the sex are in recognising an admirer, I do not believe she had any thought that Ernst was other than an artist intent upon his copy, so single-hearted was she in her own pursuits. But this could not last always. The "Village Wedding" was finished, and our heroine, after an absence of a week—during which time Ernst was inconsolable—reappeared at the Louvre, and, selecting a picture in another part of the hall, again commenced her labours. It was a landscape by Salvator Rosa, a painting calculated to call forth all her enthusiasm, and she began it with a zeal delightful to witness.

"What am I to do now?" said Ernst, despairingly. "Be near her I must—I live but in her presence. What will become of me?"

"You should paint her; then you will have her image to worship."

"Ah! would I had the right to do so—but I will not steal a portrait; I should despise myself for ever after."

"By the way, where is your Diogenes?"

"That *is* a most excellent joke. It is the only funny part of the affair. *My* Diogenes, indeed! No one after this will accuse me of *copying*."

"But what have you done with it?"

"Done with it? Nothing: I gave it to Laurent to amuse his children."

"Then I must get it from him. I will give him two pieces, much more suitable for children, for the one which he has, and preserve it for exhibition, when you are renowned."

"But that does me no good now. Let me reflect: I do not dare venture again to copy next her; she would certainly notice it."

"She would not: and that is why I admire her."

"Well, let us see, then, what I am to work at." We moved toward the spot where the girl was sitting.

"The dead Christ."

"I will not place myself there," said Ernst, emphatically. "Why will artists spend their labour on death? as if *representation* was their sole work. Believe me, it is a false idea. Life, *life* always. We have nothing to do with *dead* bodies." And he repeated his favourite quotation.

"Look on the other side."

"A sketch of Paradise." That will do. The *living* Saviour is there. This I will endeavour to transfer, and *she* shall inspire me."

A short time after this conversation I went to Havre for the purpose of taking leave of one of my relations who was about embarking for America. I was absent four days. On my return, I met Ernst standing at the entrance of our house; he expressed much satisfaction on seeing me, and appeared, I think, more cheerful than usual.

Here Franz von Herberg stopped and mused for a moment.

Messieurs, (he continued,) what I am about to

relate was told me by Ernst himself. I will proceed and take up the story from the time of my leaving for Havre, until my return to Paris—a period, I have said, of four days.

On the day of my departure, Ernst went as usual to the Louvre, and took his accustomed seat. He had really done something towards copying Tintoret's Paradise, and was certainly much improving it. I have it now in an unfinished state, and you shall see it. The girl, too, was busy with her pencil, while the very proximity made Ernst sufficiently happy. The next day Ernst resumed his seat at the usual time, but the young girl was not there. A half-hour passed and she did not come. Five minutes more—Ernst saw her walking along the gallery. His heart beat tumultuously. He could scarcely restrain his emotion. As the object of his devotion approached, he perceived that she was not accompanied by the man-servant who invariably attended her. She walked, however, rapidly forward, cast an uncertain glance around, then placed a chair for herself, and arranged for her morning's occupation. Ernst observed, however, that her countenance bore a troubled look, and that her dress was in disorder, and some parts of it seemed to have been recently soiled and draggled with mud from the street. She continued to wear both hat and shawl. This of itself would scarcely have attracted Ernst's notice, were it not for the strange appearance which the young girl exhibited. So much was he carried away by it, that, forgetting his previous resolution, he seized his pencil and commenced sketching her.

While he was thus engaged, and utterly absorbed in the occupation, the subject of his sketch rose and stepped toward him.

Ernst coloured crimson, and, like a guilty wretch, unconsciously drew aside the paper on which he was drawing.

"You were taking me?" she said.

"On my honour," cried Ernst, deeply moved, "on my honour, it was involuntary;" and he tore the paper in pieces to prove his sincerity.

"But do you desire to paint me?"

Ernst dared not raise his eyes. His first impulse was to fall at her feet and pour out his soul to her, for the tone in which she asked the question implied a willingness to grant the favour.

"Do you desire to paint me?" she repeated.

"I would ask nothing more in this world, could I have permission."

"It is granted. But you must come *now*. I can give you but *one* sitting."

"I will attend *Mademoiselle* immediately."

"Nay, I will attend *you*."

Ernst hesitated.

"*Monsieur* is losing time."

Ernst von Wolzogen was taken by surprise. What could it mean? Had he mistaken the character of his adored object? No; he could swear—No! Was it possible? Had she discovered his secret devotion, and was she therefore willing to show him this favour from a sense of pity? As yet Ernst had not presumed to look at her, but sat spell-bound.

"We lose time," she whispered softly.

Ernst started up, and, bowing low, led the way out of the gallery.

They descended the steps together, and stood on the pavement. Ernst beckoned for a carriage. His companion uttered a faint exclamation, too indistinct to be understood, and said hurriedly, "I will walk."

They proceeded on in silence. Reaching the house, the young girl followed Ernst up the staircase and into his apartment.

"Where," said she, "shall I sit?"

Ernst hastened to place his visiter; then he arranged the canvass, and deciding on what he thought the proper distance, he seized his brush.

For the first time, he now looked steadily at his companion.

She had thrown aside her hat and shawl. Her hair,

escaping from its fastening, lay in disorder over her shoulders. The face—the eyes—Ernst dropped his brush. He was terror-stricken.

"We lose time," once more she repeated.

Ernst again took up the brush; he fixed his eyes boldly on the sitter; he sat to work; he grew more and more excited; touch after touch was laid on; no point was omitted. His labour was so intense that he felt his breath shortening and his pulse throbbing as he proceeded.

"The hour has expired. I must leave you," said the girl, and she rose to depart.

"Stay—stay; in Heaven's name, stay—one instant. The eyes—the eyes—I *must* have another glance."

She turned her head; she fixed her gaze intently on Ernst for at least a minute; then waving her hand to prevent his following her, she slowly walked away.

Ernst continued at the picture the entire day, without the slightest intermission, and when evening came he laid it aside, finished. He went to bed, but he could not sleep. To use his own expression, those eyes were *burnt into him*. How would this adventure end? Would she be at the Louvre the next day? Would he ever dare address her? *Was* his visitor really the same person he had beheld so often there. She was and she was not. What could it mean?

Ernst passed the night, his brain teeming with tumultuous thoughts, and his heart beating with violence all the time. The morning dawned and found him feverish and excited. He rose and hastily dressed himself. His first impulse was to inspect the portrait. He went to his easel; he looked on the canvass. His teeth chattered; his knees knocked together.

At that instant, the woman who had charge of the room entered with his breakfast and the morning journal.

Ernst swallowed a cup of coffee. Taking up the journal, the first paragraph which met his eyes was the following.

"MELANCHOLY OCCURRENCE.—Yesterday, as Mademoiselle de Launy, only daughter of the Comte de Launy, was proceeding in her carriage to the Louvre, which she was in the habit of visiting daily, the horses took fright near the corner of the *rue de Rivoli* and the *rue Castiglione*. As the postillion endeavoured to curb them, one of the reins broke, and the horses becoming unmanageable ran furiously down the street, upsetting the carriage with great violence, by which Mademoiselle de Launy was thrown out upon the pavement and her skull fractured. She was taken up senseless, and immediately conveyed to the residence of the Comte, where every means that medical skill could suggest were resorted to, but in vain. She continued insensible, and after the lapse of one hour, life was extinct."

Ernst read no more, although the paragraph contained particulars of the beauty of the deceased, her accomplishments, her virtues: he threw down the journal. Did a shivering seize him? Was he maddened with excitement, or struck with horror? Quite the contrary. He was perfectly calm and tranquil. His own convictions were sustained and carried out: he felt a serious pleasure that *a sign had been made to him.*

The following day I returned. I found Ernst, as I have said, more cheerful than usual. Never before had I seen him so free from gloomy thoughts and fancies. To be sure, he was not gay or animated, but he never appeared more rational. His favourite author was Schiller. He felt a sympathy with any thing from his pen. As we sat together the morning in which he gave me the account I have now detailed, he repeated from Schiller's dying words, "'Now is life so clear! So much is made clear and plain!' Think you," he continued, laying his hand upon the table, "that this base matter is more enduring than spirit? I can now answer Schiller's question:

'————————————————See
The marble-tesselated floor; and there
The very walls are glittering livingly
In clearest hue and tint. The artist where?
Sure but this instant he hath laid aside
Pencil and colours!'"

I did not think it judicious to raise any discussion about a subject so delicate, although Ernst and I had been for years in the habit of canvassing each other's opinions with great freedom. Besides,—the painting. It would have been idle, were I disposed, to assert, what I by no means felt sure of myself, that it was the work of a heated and overwrought brain; that, distracted by disappointment in not meeting the object of his passionate adoration, his feverish fancy had supplied the rest. I neither affirmed nor denied what Ernst would say, but endeavoured to minister as much as I could to his prevailing cheerfulness. We continued to take our walks together; we discussed subjects of art as before; but my friend never took up his unfinished pictures ; *he never again entered the Louvre!*

"Franz, I shall never paint any more," he said to me as I was urging him to resume his labours. "I cannot," he continued, "explain to you how I feel. My devotion for Art is not lessened, nay, it is stronger in my heart than ever. I am neither moonstruck nor melancholy. What has happened to me *is natural*. But the flesh is weak. I *cannot* sit again at the easel after——"

He did not finish the sentence: he knew I understood him.

Ernst proceeded: "I must change my life. I must court an active life. I will busy myself with the practical——"

"And thy *artist*-life, O Ernst!"

"Shall still live, Franz, in my soul: it shall show itself in my deeds: they shall be humane, truthful, energetic, and so I will *create* a new picture. Behold my faith

> 'Six thousand years has death reigned tranquilly!
> Nor one corpse come to whisper those who die
> What *after* death requites us!'

No longer am *I* without assurance. This is why I am cheerful, hopeful; I believe in the '*requiter.*'"

I did not attempt to dissuade him. I could not; for I was myself convinced that Ernst was right in his decision.

His plans were not settled, but he determined first to devote a few months to travel and recreation.

The time had come when I was to lose my early friend and companion. We parted with an understanding that we should meet during the season in our native village.

Ernst decided to pass through Switzerland. It was as yet too early to cross the higher passes of the Alps with safety. But Ernst was always enthusiastic among such scenes, and loved the excitement attending them.

You doubtless remember a published account, about eighteen months ago, of a company of five persons who, attempting to cross by the pass of the St. Gothard, were overtaken by a *tourmente* near the fatal *Buco dei Calan-*

chetti, and buried under the snow. Ernst von Wolzogen was one of the party, and perished, beneath the avalanche.   .   .   .   .

.   .   .   .

There was a long pause after Von Herberg had concluded. It was broken by Vincent.

"Do you know," he said, "that story makes me feel deucedly *unsettled?* You Germans are a fearful set of fellows. What *is* the use of harrowing up one's fancies in this way? Franz, my dear boy, I mean no offence; with you it's all very natural, but it's too hard work for me; besides, my old aunt would say that it isn't good Bible doctrine. Gentlemen, you must all adjourn to my room. Franz, you shall lodge with me to-night—I have two beds, you know. I am afraid to leave you alone after such a narration. Lock that closet-door and throw away the key—g-h-r-r-r-r! It makes me shiver to think of it. *Allons, Messieurs,* I have some champagne wine and a box of real Habanas just smuggled, and, what is more, I propose to tell you a story which I heard but yesterday, and which, I hope, will help us to forget this one, so that we may sleep in peace without those *eyes*—g-h-r-r-r-r! *Allons—allons.*"

Not one of the party had stirred while Vincent was making his speech. But the spell was now broken, and, accompanied by Franz, they all descended to Vincent's room, making numerous lively demonstrations on the way. The corks flew from the champagne; pipes, meerschaums and segars were lighted, and after a reasonable period spent in discussing their merits, Vincent was called on for the story.

CHAPTER VII.

VINCENT TELLS HIS STORY.

"You all know Paul Ferval, the water-carrier?"

"Oh, certainly, we all know Paul—if that's any assistance to you."

"*Messieurs*, I beg you not to be impertinent; *au contraire*, pray be docile, and tell me when any of you saw Paul last."

After considerable serious reflection, none of the company remembered to have seen him for several weeks. It was strange; they had not thought of it before. What had become of him?

"That is what I am about to tell you. The old woman who rents the *atelier* where Paul lodged, just around the corner, in the *rue Neuve St. Médard*, has given me the whole story. It is a capital one. Our worthy Doctor Lanote is mixed up with it, and you will say the affair is very characteristic of him. But, artist-like," (bowing to Franz,) "I will begin at the beginning. *Messieurs*, please to observe silence while I give you the story of

The Water-carrier.

In a small village, a few miles from Macon, on the road to Lyons, lived—and, I trust, still lives—the widow Ferval. She had been the wife of a weaver, who, several years before the commencement of my history, selected the little village for his home, hired a small tenement, and set up his loom. It was whispered about that Ferval had, as the phrase is, seen better days. Nothing positive ever transpired, however, to confirm this notion. The weaver and his wife were both industrious—went very little among their neighbours; but, at the same time, were held in good esteem as peaceable and quiet people. They had but one child—a mere lad—when Ferval first came to the village, and who was greatly indulged both by father and mother. Young Paul Ferval grew up to be one of the finest fellows in the whole county. His voice was clear and ringing, his eye bright, his form manly, and his step full of activity. He sang a good song, he could play on half-a-dozen instruments, he knew how to cast an account and to write, and even had some taste for reading! But the worst of it was, he was taught nothing by which he might, in due time, earn an honest livelihood. He had not been put to any trade. He could not even weave; which was strange enough, as he came up under the very sound of the shuttles. In short, he never had done what one should call a day's work in his life. A very bad example did Paul Ferval unconsciously set to the youths of the village—an example which would doubtless have been gladly followed, had their fathers been like the father of Paul.

On several occasions, certain of the more substantial villagers ventured to remonstrate with the elder Ferval on the course he was pursuing with his son, and hinted that it would be much better for him to bring up the boy to some honest calling, than to permit him to be roving about the country, singing songs and playing the flute or violin. These suggestions were to little purpose. Ferval would say to his advisers: "Has my boy been guilty of any thing culpable, or any thing dishonourable? Does he frequent the wine-house? Does he keep bad company? Is he from home at unseasonable hours?" No one could assert this; and the conference would be closed by a shrug of the shoulder and a shake of the head, and a hint that the example Paul set to his companions, who were taught to labour for their living, was a very bad one, nevertheless. Ferval's father would make no reply, and so it would end. The fact is, Ferval's neighbours were right, and he was wrong. But Paul was an only child, a darling child,—and a right good child he was, dutiful and affectionate, and withal a manly fellow, —so that the father, who detested his own trade, in which, however, he was very skilful, and, being able to support his small family without difficulty, could not bear to set his bright, ardent, vivacious boy down to the back-breaking machine at which he himself toiled so faithfully.

But the evil day came at last. Ferval was taken mortally ill, and died. He left scarcely more than enough to provide for a decent burial, and his widow had to depend entirely on her daily labour for support. It was now that Paul lamented bitterly that a different course had not been pursued with him. He looked with feelings of envy on the young fellows of his own age who were already able to earn a decent support. He blamed himself for his improvidence. He knew not what to do, unless to become a labourer in the fields. Paul had another trouble, and it was a serious one.—He was in love! Fanchette Crosier was the prettiest maiden in the whole department. I won't attempt to describe her to you, gentlemen, because description is not my forte; besides, she has not been particularly described to me, and I forbear to draw on my imagination, but leave you to draw on yours. All I know is, she was confessedly without a rival the country round. Her father, Nicolas Crosier, was a stout-built, sturdy-looking old fellow, with a visage sour enough to frighten any youth who should have the audacity to offer himself as suitor for his daughter.

He had from a very small beginning got to be the *propriétaire* of a large farm, and now enjoyed himself in cultivating his own land. The girl was his only child, and, although Nicolas was at times rather severe with his daughter, his heart was really bound up in his little Fanchette, as he called her.

Nicolas Crosier was one of the persons who used to take the liberty of remonstrating with the weaver Ferval about the young Paul. After all, there was something beyond the mere desire of rendering Paul a service, and preserving the place from the evils of his example, that influenced many of those who were so ready with their advice for Paul's benefit. If the truth must be told, all the girls of the village were in love with him, and I dare not assert the pretty Fanchette was an exception—I will be frank for once, and say she was not an exception. No wonder, then, that the worthy fathers trembled at the thought of having such an idle fellow stealing in on them, and running off with the flower of their flock.

As for Paul, as I have said, he had his own troubles in this respect. *He* was in love—in love with Fanchette Crosier, and of course was in despair. In the first place, it was not possible Fanchette could ever fancy him—no, not possible. Then old Nicolas Crosier! To be sure, Paul was always so civil, so respectful, so courteous, that the old fellow could not quarrel with him. Paul had done nothing, had said nothing, had made no demonstration which approximated toward making love to Fanchette. But Nicolas Crosier was too knowing to be deceived. He kept a strict watch both on Fanchette and Paul, resolving at the proper time to give a death-blow to any hopes the latter might entertain in that quarter. I don't mean to say that the old fellow had any special objection to the youth beyond what he urged to his father—and which certainly was very proper—but he had made up his mind, after the death of the elder Ferval, to put a stop to Paul's coming to the house as soon as he decently could do so. It will be seen that Paul himself brought on the crisis. He could endure the suspense no longer. So one morning he goes to the house of Nicolas Crosier, which was situated a little out of the village, determined to seek an interview with the old man, and have his destiny settled.

Nicolas was seated on the little portico which skirted the front of his house, and which overlooked his garden and his meadow. He read Paul's errand in his face, and was glad enough that the wished-for opportunity had come. He saluted the young man civilly, and bade him take a seat. The latter was too much agitated to sit down, but told his errand at once.

"You want to marry Fanchette?"

"With your permission," said Paul, modestly.

"What would you do with her?" asked Nicolas, gruffly.

Paul hesitated: he was not taken by surprise, for he knew the question would be put to him; but now that it *was* put, he felt the force of it more than when he was considering it by himself.

"What would you do with her, eh?"

"I would work from morning to night, and she should want for nothing," said Paul, resolutely.

"These are fine words, and you are doubtless a very fine fellow," said Nicolas, ironically; "but tell me, Paul Ferval, are you really such an *imbecile* as to suppose me willing to throw away Fanchette on a lazy, idle vagabond —one who never earned the salt in his soup, and now that his father is dead, is seeking to be supported by a father-in-law."

Paul swallowed the hard words with difficulty; the insinuation of seeking a support cut him to the quick; at the same time he could not deny but that it would be very natural for any one to view his conduct in just that light.

After a moment's hesitation, he replied, "I do not wonder you have these suspicions, but you wrong me. I do not want Fanchette until I prove to you I am able to support her."

"*Cela est fort beau ; mais à quoi diable cela revient-il ?*" asked old Nicolas, sneeringly. "What's that to the purpose?—why do you come to me *now?*"

"Because," said Paul, with a despairing energy, "if I had from you the slightest assurance that Fanchette might one day be mine, it would give me courage to accomplish every thing, and this, *Monsieur* Crosier, is why I come now."

There was something in Paul's resolute tone which touched a similar chord in the old man. Besides, Paul sought no present advantage: he was content to put off the day: there was one point gained. Nicolas Crosier considered a while, and then he said—" Paul, your father was a worthy, industrious man; your mother is a most excellent woman; *you*, at present, are a miserable, worthless do-nothing. You say you are resolved to turn about, go to work, and make something of yourself: I don't believe you ever will; but I am not the one to discourage a man who wants to do better. Fanchette is but sixteen. She sha'n't marry *any body* with my consent these three years. Now, look you, if you can come to me in three years, and say—'Nicolas Crosier, I have earned money to buy some land' (I don't care how little) 'and I have saved money to build a cottage,' (let it be ever so small,) 'and I want Fanchette,'—*sacré bleu!* you shall have Fanchette—that is, if the girl is fool enough to say she likes you, which I very much doubt; and here is my hand on it."

Paul seized the offered hand, and gave it such a grasp that it brought the tears into old Crosier's eyes.

"I don't ask for better terms—I have no right to ask for better. Ten thousand thousand thanks."

"*Brisons là-dessus,*" said the old man, hastily; "go in if you like and see my wife, perhaps Fanchette is with her; then be off with yourself. No more love-making—do you understand?—for three years."

It is very doubtful if Paul heard the last part of the sentence, as he was already in the house, seated between Fanchette and her mother. He told the latter (with whom, by the way, his handsome address and pleasing manners had made him a favourite) the result of the late interview, and improved the short time that was allowed him, I have no doubt, to the best advantage. A three years' banishment is certainly a formidable obstacle even for "true love:" but Fanchette had no fears; her mother had no fears; so, with many words of encouragement, Paul took his leave. Old Nicolas Crosier nodded carelessly to him as he passed out; and it was not till Paul had entered his mother's cottage that his heart sank within him at what he was to undertake. But his resolution was fixed. He briefly informed his mother what had occurred, and begged her to grant him–her blessing, and let him set out the next morning to seek his fortunes. It was a grievous struggle for the poor woman, but her son's reasoning finally prevailed, and the next day Paul departed on foot from his native village. A knapsack swung over his shoulders contained his clothing, and the sum of twenty-six francs, which the widow had carefully saved for some unforeseen emergency, was safely deposited in his pocket. It was a long time before Paul consented to take the money, for the sneers of old Nicolas Crosier were still tingling in his ears; his mother, however, who knew how much he might stand in need of it, forced the silver into his hand, and, throwing her arms around his neck, she embraced him tenderly and commended him to the keeping of the Holy Virgin and all the Saints. Paul sobbed with grief, in spite of himself, as he trudged slowly away. But, as he got out into the open country, the fresh fields and the pleasant prospect inspired him, while the thought of the stake for which he was venturing soon restored all his natural courage and determination.

His journey contained no adventures. He was kindly entertained by the inhabitants as he passed along, all of whom were delighted with Paul's open, easy manner, and pleasant, cheerful countenance.

It was about noon that our adventurer found himself, after some days' travel, within sight of Paris. His purse still bore the weight of his twenty-six francs, for so hospitably had Paul been treated upon the road that he found no occasion to lighten it.

His heart beat with excitement as he beheld the gay city where all his hopes were centred. He was very sanguine. If he had been received as a brother by the peasants by the wayside—some of whom were nearly as poor as himself—what good fortune must now attend him? what might he not expect from the rich and the powerful? Poor Paul had yet to learn the lesson that kindnesses and charities spring from the humble and the lowly, not from the opulent and the great.

As he advanced into the city, he reached the Italian Boulevards. They were thronged, as usual, with a glittering crowd, intent on pleasure and pastime. Paul gazed wildly around: the stream swept by him in a never-ending current. He put a civil question to one of the passers-by; it was answered by a shrug and a stare. Gay sights filled his eyes, lively voices were sounding in his ears, brilliant equipages swept rapidly along, and the shops and *cafés* and saloons bewildered him with their dazzling glare.

Paul's heart sank within him. He thought of his native village, of his mother's cottage, and his courage failed him. The real state of things flashed, by a kind of prescience, on his mind. Where was he to go? what was he to do? He felt that he had no power to interrupt the passing pageant with the story of his wants or of his aspirations, so he stood oppressed and dejected, till, finding himself continually jostled by the crowd, he proceeded down the Boulevards in the direction of the river.

It was doubtless by that same instinct which leads the miserable to seek the companionship of the distressed, that Paul found himself, as the day began to wane, on this side of the Seine, and in the dirty quarter of the *rue Neuve St. Médard*. He was weary, hungry, and dispirited. The purse containing his twenty-six francs was still safe in his pocket; but he dreaded to make the first inroad upon it.

As he stood leaning against the doorway of one of the ponderous buildings, irresolute what course to take, a little, fat old woman, fifty or sixty years old, with a large wool mattrass on her head, turned into the court-yard. She did not see Paul, who at that moment had advanced a step directly in her path. He, poor fellow, was too much taken up with his own situation to notice her. At the moment of contact Paul made an awkward endeavour to avoid the collision. It resulted in making matters worse. The mattrass was not only thrown down, but the little, fat old woman unfortunately lost her balance, and rolled into the mud.

Paul hastened to her assistance, but was greeted by a storm of abuse for his carelessness in intercepting the passage-way. They were the first words which had been addressed to him since he placed his foot within the city, and it was music to his ear to hear them. He raised the old woman in spite of her clamour, took up the mattrass with alacrity, and insisted on carrying it to the top of the house, where was situated the small *atelier* in which she performed her labours of *cardeuse*.

By the time Paul had mounted *au sixième*, the wrath of "Old Mannette," the little fat woman,—we may as well call her by name, at least, the only one by which she is known,—was very considerably abated; and when, having been directed into what room to go, Paul put down the mattrass and again asked pardon for his awkwardness, Old Mannette's feelings took quite the contrary turn, and she apologized with much volubility for her own rudeness. The old and ill-formed are especially gratified by the attentions of the young and handsome. It certainly did not diminish the force of her protestations when she saw that it was a fine, manly-looking fellow who was showing her so much civility.

The result of this adventure was very satisfactory, all things considered. The sight of Paul's knapsack naturally called forth an inquiry from the old woman, and it ended in Paul's telling her his whole story. Just think of it! Paul Ferval making a *confidante* of Mannette, the old *cardeuse!* A heavy falling off from the bright anticipations of the morning. For all that, Paul was happy enough to find any body to whom he could talk, and of whom he might ask questions.

Old Mannette, after all, was not the worst adviser

Paul could have had. She was really a kind-hearted, sensible creature, who understood the ways of the town, having been left at an early age to take care of herself. Of course, she had never married; for nobody would think of such a dwarfish, ill-formed thing for a wife. She had now lived many years in that house, and pursued her vocation unremittingly from day to day. Such was the person who, before she was aware of it, had taken a strong interest in Paul's fortunes. And Paul—he was no longer the lonely, miserable, isolated wretch, surrounded by the gay throng of the Boulevard. Seated on a pile of mattrasses in the little dark *atelier* of Old Mannette, he was as light-hearted and happy as was possible.

"I tell you what you shall do, Master Paul: I have a little room which joins this, not much larger than a closet, I admit, but it will answer until you can afford something better. You shall have as many mattrasses as you like, and I can manage to make up the bed for you from my own store. *Au reste*, you shall breakfast with me, paying only your share, and for dinner I have always an abundance of soup and excellent *bouilli*. But that, after all, is nothing," continued the now enthusiastic old woman. "What can be thought of for you?" And she put a multitude of questions to Paul as to the extent of his capacities.

The poor fellow made but a sorry figure while going through this catechism.

"I have it at last," cried Old Mannette with delight; "you shall become a *water-carrier*. The old carrier left the 'route' only yesterday; to be sure, he could not make a living out of it, but then he has not half your activity, I am sure he has not. To-morrow you shall begin: you must purchase your cans early in the morning, and I will go with you and show you the way."

Old Mannette now went into a full explanation of the labours and duties of a water-carrier; and, although she admitted he would have a very unpromising route, still she was persuaded Paul could make something out of it.

Our hero went to bed with a light heart. Doubtless, the thought which is so well expressed in our English lines came into his head:

" 'Tis the poor man alone,
When he hears the poor moan,
A mite of his morsel will give.
Well-a-day."

The next morning Paul rose bright and early. Not now with reluctance was his purse drawn forth. He was paying money for the implements of his trade, and he counted it out cheerfully. Soon he became familiar with the mysteries of his profession, and settled into its routine. It *was* hard, even for Paul, to make a *sous* over and above his daily expenses. Old Mannette had not done justice to the former water-carrier. He had abandoned the route after a very faithful trial. But Paul was not to be discouraged. He *did* gain a little. By degrees he extended his trips as he gained greater facility in serving; he also made inquiries about the business in the other parts of the town, and discovered that to be in a position to lay by any money, he should be possessed of a horse and cart.

"What shall I do," he said to Old Mannette, "for a horse and cart? I *must* have a horse and cart. I am so familiar with the work now, that I could soon change my situation with a horse and cart."

"Take time, Paul, take time," Old Mannette replied; "do not fret too much about it. I know a wheelwright near by, who will, I am sure, let you have a second-hand cart on credit, if you can only buy the horse."

Paul set to with more zeal than ever, and by degrees his purse grew a little heavier, and his heart proportionably light.

The weather at length began to be cheerless and forbidding. The winter, which is always disagreeable in Paris, was unusually severe, and Paul overtasked himself in performing what were his accustomed summer duties. In one very severe storm Paul was more than usually exposed, and he continued to labour till a late hour. He came home shivering, and, without taking proper pains to dry himself, he went to bed. He awoke in the middle of the night with a burning fever. He tossed and tumbled about till morning, and then endeavoured to rise. His limbs refused to sustain him, and he sank almost fainting on his bed. After a few moments he again attempted to raise himself, but the room seemed to whirl round, and he grew so giddy that he was forced once more to throw himself down. Shortly after, Old Mannette, having prepared breakfast, knocked at his room, surprised that Paul was not already stirring. She was answered by a voice so uncertain in its character that she pushed open the door. She hastened to his bedside, and, seeing the poor fellow so ill, could not help expressing her lamentations. Paul had never been indisposed a day in his life before, and had not the slightest idea of a prolonged sickness. Much, therefore, as he was suffering, he assured the kind-hearted old woman that he should be quite well in a few hours, and asked her to see one of the *ouvriers* below, and arrange for the performance of that day's labour. Old Mannette was familiar with disease; for at one time she had been a nurse: her experienced eye detected the presence of a violent fever, and she was satisfied that it would confine Paul many days to his bed. She did not discourage him, however, but endeavoured to persuade him to remove to her own room, which could boast of the luxury of a fireplace, although it is more than probable it had never been used. Paul would not consent to change, and Old Mannette, after preparing a ptisan and placing it near him, and recommending him to keep very quiet, went out to attend to her occupations. When she returned, she found Paul much worse. Indeed, Paul himself began to think something serious was fastened on him. The very loss of time was to him a mortal blow. Six months had elapsed, and he had certainly gained very little toward the consummation of his plans. Here was a "stay" which Paul had never calculated on—had never dreamed of. Old Mannette read his thoughts, and hastened to comfort him. First, she insisted on his removing into the next room; then she purchased fagots and made a fire, without letting Paul know it was not her daily custom to do so. What with the excitement of the change, and of the pleasant blaze upon the hearth, her patient fancied himself much better, while Old Mannette had not the heart to tell him she knew he was hourly growing worse.

He passed a miserable night, so that Old Mannette was much alarmed at his appearance the following morning. However, she said but little to Paul, who, now almost despairing, lay, with throbbing pulse and suffused eyes, in a state bordering on entire apathy. Immediately after her breakfast she hurried away to "dear Dr. Lanote," Old Mannette's favourite among the medical faculty. You all know the Doctor, his oddities, his eccentricities, his abrupt manner, and his kind heart. If the worthy man stood so high in Mannette's critical estimation, the latter was no less esteemed with him. She had nursed under the Doctor, first and last, at least twenty years, and, as he always bore witness, she had never swerved from his orders or volunteered any thing out of place. A reliable nurse is a valuable assistant to the physician, and whenever Dr. Lanote was called to a patient and found Mannette in charge, he did not hesitate to express a peculiar satisfaction.

To Dr. Lanote Old Mannette hurried. She had to cross the river, and take a long walk on the other side of the Seine, but she got to his house before he had left, and waited for her turn to see him.

"Ah, Mannette, is it you? Sit down. You have had a long walk this morning. I hope you have come to say you are going to try your hand as nurse again? Nowadays it's a novelty to find a nurse that keeps to her duty. Do you remember how we carried through Monsieur Gaudelet?"

"Can I ever forget it?" said the old woman with vivacity; "and how the rich merchant offered you his purse, and you bade him hand it to Mannette, saying that the nurse had done more than yourself towards his recovery. It was a proud day for me; and that was the reason I could not receive the money. I felt more than paid by what you said of me, and it seemed as if I should be robbed of that reward had I taken it."

"Ah, Mannette, you are an old fool like myself. But I must despatch these patients. What is to be done for you?"

Although Dr. Lanote was really in a hurry that morning, he sat patiently and heard Mannette tell the whole story of Paul Ferval from beginning to end. He betrayed no particular interest in the narration, and when it was concluded he said, "Very well, Mannette. Now go and attend to your mattrasses, I will give the youngster a call;" and the old woman went off as well contented as if she had received every assurance in the world in relation to her *protégé*.

It was about one o'clock in the afternoon. With Paul the hours had passed heavily. From a continual restlessness he gradually sunk into a stupor, against which he vainly endeavoured to struggle. He was partially roused from it by a friendly shake of the shoulder, while an abrupt but not harsh voice exclaimed, "Well, what are you doing here?"

Paul opened his eyes and saw standing over him a little, inquisitive-looking, bright-eyed old fellow, who was regarding him with an expression of curiosity and interest.

"Well, *mon enfant*, what do you think of me?"

As Paul knew nothing of Old Mannette's expedition, he had not the slightest idea that the visit was from a physician. He said nothing, but stared wildly at the intruder.

"What is the matter with you?" said Dr. Lanote.

"I don't know," replied Paul.

Dr. Lanote proceeded to examine his symptoms very carefully.

"Are you a medical man?" inquired Paul.

"Yes."

"I don't want one."

"Why not?"

"I am better without."

"You speak more truth than you imagine, my poor fellow," muttered Dr. Lanote to himself.

"Besides," continued Paul, "I have no means of paying for visits."

"That is not true," said the doctor, bluntly. "You have a bag containing five-franc pieces some where about the room."

"Wretch!" cried Paul, in an excited tone, "would you rob me?"

"No," said the doctor, dryly, evidently not relishing

being mistaken for a *voleur*. "I would cure you, that's all—and to do that, I must rouse you; and I think I have partially succeeded. Where's Mannette?"

"I cannot tell," said Paul, who began to think he was in a dream.

At that moment the old woman's step was heard on the staircase, and the next, she made her appearance in the room. Dr. Lanote took her aside.

"Over-exertion of body and mind,"—he whispered,—"grief — care — disappointment — *cerebral* — *typhus*. He must be *nursed*—you understand."

He continued his suggestions much in the same manner, dropping a word here and there, to represent a whole sentence, which was doubtless coherent enough in his mind, and which Old Mannette seemed to understand perfectly.

"I will be here in the morning," said Dr. Lanote; then turning to Paul, he bade him take courage, and took his leave.

"Is it possible that there are no remedies for such a disease?" said Dr. Lanote to himself as he stepped slowly down the staircase. "I am convinced that neither the antiphlogistic, nor the stimulant, nor the tonic, nor the derivative, method of treatment is of any avail. I *dare not* follow any of them. To what, then, am I reduced? To the EXPECTANT! Just what that sensible American* declared after a practice of nearly half-a-century, to wit, 'that we had better leave the disease to cure itself, as remedies, especially powerful ones, are more likely to do harm than good.' Well, well, the boy has a stout frame, and by carefully watching—but his courage is gone—his courage is gone—there's the rub;" and Dr. Lanote got into his carriage and drove away.

The Doctor called daily, sometimes twice a day, while the fever gradually crept through Paul's system, and approached the crisis. He had taken no medicine. Dr. Lanote would prescribe nothing, except, perhaps, a little barley-water, weak lemonade, or something of the sort.

Notwithstanding Old Mannette was as economical as she could be, it was necessary to make some trifling purchases which she had no means of supplying. Paul had at first resolutely resisted any encroachments on his treasure. One day Dr. Lanote came in and recommended her to procure two or three little articles which were really necessary for his comfort. Old Mannette looked mournfully at Paul. "You hear," she said, "what the good Doctor advises?"

* Dr. Lanote doubtless referred to Dr. Nathan Smith, unquestionably the most eminent physician this country has ever furnished, and who adopted and introduced a new method of treatment in typhus fever.

"I must do without them," said he, in as decided a tone as his weak voice would permit.

"A miser, and so young!" cried Dr. Lanote a little sharply.

"I hoarded my money to buy a horse and cart," answered Paul, bitterly.

A compression of the lips and a slight tremulous movement of the muscles of the mouth could be perceived; but the Doctor manifested no other sign that he had heard a word that Paul was saying.

As for Paul, he now submitted to his fate, because he could no longer resist it. His hopes were fled, every one of them, and he really did not care what became of him.

By degrees his purse grew lighter and lighter, for Dr. Lanote *would* have his own way, and Paul ceased to give a thought on the subject.

The Doctor continued unremitting in his visits, and kept the strictest watch of every symptom, so that he might check at once any of those intercurrent affections which are so apt to appear in the disease, should any be manifest.

The fever at last had spent its force, and the crisis approached. The principal danger was to be apprehended from Paul's utter despondency. I should not say despondency. He had reached a lower point, for he had ceased to despond. If he had any wish at all, it was that he might escape from the world. Poor Paul!

It was near evening. Paul had been sick fourteen days, and the crisis of the fever had come. Dr. Lanote stood by our hero's bedside with a perplexed aspect. At last he said to him, "*Mon enfant*, you must bestir yourself—pass but a good night, and to-morrow you will be better."

"I do not want to be better," said Paul, faintly—"I want to go; it will be soon, I hope."

"Ah, very well," said the Doctor, "if I can be of any service to you, command me. I will see to any thing you intrust to me."

Paul made no reply, but whispered, in a low tone, "Fanchette."

"Fanchette?" cried Dr. Lanote. "What Fanchette? Is it the Fanchette who is soon to be married to Jean Grilliet."

Paul opened his eyes very wide, notwithstanding his weak state: an electric shock had been administered.

"What do you say?" he asked with considerable energy.

"I spoke of Fanchette Crosier, the lass Jean Grilliet is about to marry, and a nice fellow he is too. I only hope the girl is half as good as he is. But why do you speak of Fanchette—is she any thing to you?"

"Fanchette Crosier to be married!" exclaimed Paul, endeavouring to raise himself.

"Certainly—why not, *mon enfant?* These young girls all have their day, and so do the boys as well. If you will but bestir yourself, your own turn will come by and by. I know a pretty little mischievous creature that I will recommend for you, if you will only promise to behave yourself, worth a dozen of this Fanchette, if you ever cared for her."

The perspiration streamed from Paul's face; this time he was *completely* electrified. He looked sharply at Dr. Lanote, who stood the very personification of innocence and simplicity.

"It is false," he finally exclaimed: "I *won't* believe a word of it. If I thought it were possible——"

"Just make a journey home and see for yourself," interrupted the Doctor, "and if I have misinformed you, I will pay the whole expense of it."

"Then I have my mother to LIVE for," said our hero; and, worn out by the natural vehemence of his feelings, he sank exhausted on his pillow, and in a few minutes was fast asleep.

"Done—*c'est fini* "—muttered Dr. Lanote emphatically.

"What is done?—*Mon Dieu!*—What is finished? Dear Doctor—dear Doctor Lanote!" cried Old Mannette in an excited tone.

"Nurse," said Dr. Lanote, "I am surprised to see you forget yourself. In former times you would not have been guilty of such imprudence. You perceive that the boy slumbers naturally. It is the only thing which will save him. Keep every thing quiet, every thing comfortable. Let him sleep this way through the night, and he is safe.

"An excusable artifice," muttered the Doctor to himself; "it *touched* his vitality—case for record— will look in early;" and with another glance at the slumbering Paul, and another nod of satisfaction, the Doctor hurried away.

It happened precisely as Dr. Lanote had predicted. Paul slumbered soundly all night, while Old Mannette never left his side, and the next morning he opened his eyes very weak and very helpless, but really a new man.

Old Mannette perceived at once the happy change. She would not permit Paul to speak a word, but whispered to him that he was rapidly getting better! The latter endeavoured to collect his senses. At length what the Doctor had told him of Fanchette flashed upon him. He groaned aloud—he could not help it: then he asked himself, "Could it be true?" and then he felt an impatient desire to get well, and satisfy himself on the point. At this juncture, Dr. Lanote came gently into the room. Approaching Paul's bed, he took his hand and said cheerfully, "Now, my *bon enfant*, you have only to keep quiet and get well, and I will see what can be done for you."

"It won't do to undeceive him yet," he said to himself; "we must wait till he has more strength."

Although Paul had at first taken a great antipathy to the Doctor, he had already begun to experience a change in his feelings towards him. He even endeavoured to return the pressure of the Doctor's hand, and was about expressing some words of gratitude, when the latter prevented him from speaking.

"Not a word now—in a few days you may talk as much as you like:" and after giving further directions to Old Mannette, he directed her in a whisper not to spare the few remaining francs which Paul might have left, but to be sure and procure certain little delicacies which he was even so particular as to name to her.

Things now went on well enough: to be sure, Paul's money was all gone; not a *sous*, not a *centime*, was left in the purse his mother had given him; indeed, for several days Dr. Lanote had himself supplied all the desired superfluities. But Paul himself was gaining rapidly: after a while he could sit up a little; then he could walk a few times across the room; at length he could dress himself. He began to be very impatient to get out and breathe the air; but the Doctor restrained him, and Paul was too grateful to be disobedient. He was, however, filled with but one thought: it was to go back to his native village and satisfy himself of the truth or falsehood of the story that Fanchette had really given herself to a rival. The stronger Paul grew, the less he was inclined to credit the tale, and therefore the more desponding he became with regard to his own prospects. A singular paradox truly, but so it was, and such is human nature

It was now early in the spring, and on one pleasant day Dr. Lanote called in somewhat later than usual, and bade Paul equip himself warmly, and he would promise him a drive. Old Mannette bustled about in high spirits—indeed, with a glee that seemed rather extravagant, and which was by no means in accordance with Paul's depressed feelings. The latter, however, was soon ready, and the three now slowly descended to the street.

At the entrance to the court-yard there was a horse and cart, while a smart, active-looking young fellow stood on the latter, as if waiting for orders.

"It is the new *water-carrier*," whispered Old Mannette to Paul. Paul looked at him with a melancholy expression, and was about to turn away, when the man jumped lightly from the cart, and touching his hat, said, "Is this Monsieur Paul Ferval?"

"Paul Ferval is my name," said our hero.

"I have brought round the new cart you ordered some time since; it should have been here yesterday, but it was not quite finished. Your horse feels well this morning—he has not been used lately. He is in excellent condition for work—that you may depend on."

Paul Ferval was thunderstruck. He could not say a word, but stared first at the man, then at Dr. Lanote, and finally at Old Mannette. The doctor was the first to speak. "You may drive into the court-yard," he said to the man, "and wait till we come back. Come, Paul, I have no time to lose—get in."

The fresh air and the pleasant sun and the agreeable change had a sensible effect on Paul's feelings; but when Dr. Lanote remarked very gravely that he believed he had a mistake to correct; he had ascertained that *the* Fanchette who was to be married to Jean Grilliet, was not Fanchette *Crosier*—Paul's Fanchette—but doubtless some other Fanchette, and so forth; and when he added further, that, instead of the wager which he had ventured to make of Paul's expenses home to ascertain the fact, he thought he would substitute a good horse and cart and equipments, which he had that morning delivered, —Paul actually threw his arms around the good Doctor and embraced him, frantic with happiness.

The rest you can all guess. Paul was soon strong

enough—he went to work—he enlarged his business—he was lucky in every thing he did; he was the most successful water-carrier in all Paris. *Bravo, Paul Ferval!*

Paul kept his three years' truce religiously. I won't say, in all that time he heard nothing from Fanchette Crosier. I am inclined to think the little baggage knew just how Paul was getting on from one month to another after he began with that horse and cart.

Well, the three years were up, and Paul had accumulated enough, certainly, to come within the moderate limits set by old Nicolas Crosier.

Yes, the three years were up, and Paul had returned his native village and made glad the heart of his good, fond mother.

The next morning, after having equipped himself in his best, and received his mother's caresses and compliments, he left the cottage and took the road to Nicolas Crosier's.

It was a pleasant summer's day, and the old fellow sat after dinner on the same balcony, and in the same chair, and precisely on the same spot, where he was seated three years before, when he made the compact with Paul and relieved himself of the handsome vagabond, as he used to call him. Nicolas had altered but little in appearance, in habit, or in disposition, so far as one could see, unless to become a little more arbitrary, a little more sedentary,

"Paul Ferval is my name," said our hero.

"I have brought round the new cart you ordered some time since; it should have been here yesterday, but it was not quite finished. Your horse feels well this morning—he has not been used lately. He is in excellent condition for work—that you may depend on."

Paul Ferval was thunderstruck. He could not say a word, but stared first at the man, then at Dr. Lanote, and finally at Old Mannette. The doctor was the first to speak. "You may drive into the court-yard," he said to the man, "and wait till we come back. Come, Paul, I have no time to lose—get in."

The fresh air and the pleasant sun and the agreeable change had a sensible effect on Paul's feelings; but when Dr. Lanote remarked very gravely that he believed he had a mistake to correct; he had ascertained that *the* Fanchette who was to be married to Jean Grilliet, was not Fanchette *Crosier*—Paul's Fanchette—but doubtless some other Fanchette, and so forth; and when he added further, that, instead of the wager which he had ventured to make of Paul's expenses home to ascertain the fact, he thought he would substitute a good horse and cart and equipments, which he had that morning delivered,—Paul actually threw his arms around the good Doctor and embraced him, frantic with happiness.

The rest you can all guess. Paul was soon strong

enough—he went to work—he enlarged his business—he was lucky in every thing he did; he was the most successful water-carrier in all Paris. *Bravo, Paul Ferval!*

Paul kept his three years' truce religiously. I won't say, in all that time he heard nothing from Fanchette Crosier. I am inclined to think the little baggage knew just how Paul was getting on from one month to another after he began with that horse and cart.

Well, the three years were up, and Paul had accumulated enough, certainly, to come within the moderate limits set by old Nicolas Crosier.

Yes, the three years were up, and Paul had returned his native village and made glad the heart of his good, fond mother.

The next morning, after having equipped himself in his best, and received his mother's caresses and compliments, he left the cottage and took the road to Nicolas Crosier's.

It was a pleasant summer's day, and the old fellow sat after dinner on the same balcony, and in the same chair, and precisely on the same spot, where he was seated three years before, when he made the compact with Paul and relieved himself of the handsome vagabond, as he used to call him. Nicolas had altered but little in appearance, in habit, or in disposition, so far as one could see, unless to become a little more arbitrary, a little more sedentary, and a little more gray. On the contrary, Paul had changed wonderfully. His frame was stouter, his shoulders were broader, his form larger and more manly. Besides, he had cultivated, or rather left uncultivated, his beard and whiskers and moustache, after the mode called in Paris "*inculte*," and was really a formidable fellow to look at. He marched with a firm step toward Nicolas Crosier.

"*Bon jour, Monsieur Nicolas Crosier*," said Paul, in a firm, strong voice.

Nicolas rubbed his eyes, but he did not recognise the stranger.

"*Bon jour*," he replied.

"I understand you desire to dispose of a part of your farm," said Paul. "If so, I should like to become the purchaser."

"*Diable*," growled Nicolas Crosier. "And *I* should like to know who has been putting such nonsense into your head."

"I want to build a neat little cottage," continued Paul, without heeding what was said, "and it strikes me I could not be better suited than hereabouts."

Nicolas Crosier rose slowly to his feet; something in the tone and manner of the stranger was familiar to him—something, too, he seemed to recollect about land, a cottage, and Paul Ferval. He came close to Paul —he recognised him. What he would have done by way of further demonstration I am unable to say, for at that moment out ran both wife and daughter, and such a scene as there was, and such fools as they all made of themselves—according to old Nicolas, who stood waiting to put in a word, but could get no opportunity—it would be quite impossible to describe.

After a time, however, the excitement began to subside, and Paul, taking his purse from his pocket—the same purse his mother had pressed on him—now well filled with gold-pieces—handed it to Nicolas Crosier, saying, "Is this sufficient? have I performed my part of the contract?"

"*Sacré bleu!* yes, and you shall see if I will perform mine. Here, Fanchette—come here. But, perhaps," said Nicolas, stopping suddenly short and trying to assume a serious expression, "perhaps Fanchette won't *ratify*—ha —ha—ha! You know I was not to interfere with her. —Fanchette, you little witch, what do you say?—ha— ha—ha!"

What Fanchette said, and so forth, and so forth, and so forth, you may judge for yourselves, *Messieurs*, when I tell you that the wedding took place last Tuesday, and Old Mannette, who of course was sent for on the occasion, returned to town yesterday, from whom I have had the whole tale.

"And an excellent one it is," shouted all present.

"Let us fill and drink the health of Paul and his pretty wife—long life to them!"

The company broke up in great glee. Laughing, talking, singing, and making other lively demonstrations, they dispersed to their several apartments.

Nobody thought of Ernst von Wolzogen and his picture!

CHAPTER VIII.

MORNINGS AT LA MORGUE.

A MORNING at *La Morgue* is hardly as agreeable as a day at the Louvre, yet it is not without a certain fascination. Let but the influence once fasten on you, and it will be very hard to shake it off. At one period I confess it was to me almost irresistible, and I shudder sometimes, when I recollect how punctually every morning, at the same hour, I took my place on one side of that fearful room—not for the purpose of inspecting the bodies of the suicides, (I rarely turned to look at them,) but to regard the countenances of the anxious ones who came to realize the worst, or to take hope till the morrow. Literally, there are no spectators in that dismal solitude —if we except an occasional visit from the foreign sight-hunter, who comes in charge of a valet, and passes in and out and away to the " next place." In London or in New York, an establishment so public would be thronged with persons eager to gratify a prurient curiosity. Not so in Paris. The French possess a sensibility so refined—it may be called a species of delicacy—that they cannot enjoy such a spectacle, can scarcely endure it: and if the tourist will bring the subject to mind, he will find that while his guide pointed out the entrance, he himself declined going into the apartment.

I know not how it happened, but, as I have remarked, the habit of visiting this spot every morning was fastened on me. Never shall I forget some of the faces I encountered there. One image is impressed on me indelibly; it is that of a woman of middle age, with a very pale face, and having the appearance of one struggling with some wearing sorrow, who for two weeks in succession came in daily, and, walking painfully up to the partition, looked intently through the lattice-work, and turned and went away. I never before felt so strong an impulse to accost a person, without yielding to it. Indeed, I had resolved to speak to her on the morning of the fifteenth day, but she did not come, and I never saw her again. Who was she? did her fears prove groundless? what became of her? An old man I remember to have seen—a very old man, feeble and decrepit, who came once only, looked at the dead, shook his head despairingly, and tottered away: I know not if he discovered the object of his search. Young girls who had quarrelled with their lovers, and lovers who in moments of jealousy had been cruel to their sweethearts, would look anxiously in, and generally with relieved spirits pass out, almost smilingly, resolving no doubt to make all up before night should again tempt to suicide. Another incident I cannot omit, although it is impossible to recall it without a dreadful pang. One morning a pretty fair-haired child, not more than four years old, came running in, and clasping the wooden bar with one hand, pointed with her little finger through the opening, and with a tone of innocent curiosity said, "There's mamma!" The same moment two or three rushed in, and, seizing the unconscious orphan, carried her hastily away. She had wandered after some of the family, and heard enough as they came from the fatal place to lead her to suppose her lost mamma was there, and so she ran to see. What could be the circumstances so untoward, that even the child could not bind the mother to life?

A long chapter might be written of the occurrences at my singular rendezvous, but I had no design of alluding to any of them: they naturally come to mind, and I as naturally speak of them in connection with what I am now going to relate.

Before the winter was fairly upon us, I resolved to spend it in the south of Europe. Partridge, much as he desired to accompany me, would not break in on his settled plans. He was quite right; but as our professions were to be different, I had 'not so good reasons as he for remaining in Paris. Accordingly I left for Italy. In this way, I got rid of the horrible nightmare impulse to which I have alluded, and although I returned the following season I never again entered *La Morgue*. . .

. . . .

It was in the summer when I came back. The foliage was deep and green, and in the *Jardin des Plants*, which was near my quarters, the various flowers and shrubs and trees filled the atmosphere with fragrance, and tempted us to frequent strolls along its avenues.

"Come with me at six o'clock," said my friend Partridge, "and you shall see an apparition."

"Where?"

"I will not tell you till we are on the spot."

"I will go, but hope the place is an agreeable one." Just then, I know not why, I thought of *La Morgue*, and shuddered.

"The most agreeable in all Paris."

This conversation took place in the Hospital, just as we were finishing our morning occupation of following Louis through the fever wards. Partridge was once more my room-mate, having, as I have said, remained behind during my late tour, to devote himself more entirely to his medical pursuits, while I, beginning to tire of the lectures of Broussais, and the teachings of Majendie, yielded to the temptation and ran away from both;

and, even now that I had returned, was induced every day to slip across to the *rue Vivienne*, where were staying some fascinating strangers, whose acquaintance I had made *en route*, and who had begun to engross me too much for any steady progress in my studies; at least, so thought Partridge, who shook his head and said it would not do for a student to cross the Seine— he ought to stay in his own *quartier*—that I had too much recreation as it was—I should forget the little I knew—and as for the *rue Vivienne*, and the *Boulevart des Italiens*, the *rue de la Paix*, &c., I must break off all such associations or be read out of the community. I was glad, therefore, to appease my friend by consenting to go with him—I knew not where—and see an apparition.

Accordingly, a few minutes before six, we started together on the strange adventure. We passed down the street which leads to the *Jardin des Plants*, and, entering through the main avenue, walked nearly its entire length, when my companion turned into a narrow path, almost concealed by the foliage, which brought us into a small open space. Here he motioned me to stop, and, pointing to a rustic bench, we both sat down. At the same moment, the chimes from a neighbouring chapel pealed the hour of six, and while I was still listening to them, my friend seized my arm and exclaimed in a whisper, "Look!"

I cast my eyes across to the other side, and beheld a figure advancing slowly toward us. It was that of a young girl, in appearance scarcely seventeen. Her form was light and graceful, simply draped in a loose robe of white muslin. On her head she wore a straw hat, in which were placed conspicuously a bunch of fresh spring blossoms. The gloves and mantelet seemed to have been forgotten. Her demeanour was one of gentleness and modesty. She cast her eyes around as if expecting to meet a companion, and then quietly sat down on a rude seat not very far from where we were. I remained for ten minutes patiently waiting a demonstration of some kind, either from my companion or the strange appearance near us. But now I began to yield to the influence of the scene. The sun was declining, and cast a mellow and saddening light over the various objects around. Gradually, as I gazed on the motionless form of the maiden, I felt impressed with awe, which was heightened by the solemn manner of my friend, who appeared as much under the charm as myself. At length I whispered to him, "For Heaven's sake, tell me what does all this mean?" A low "Hush," with an expressive gesture to enforce quiet, was the only response. I made no further attempt to interrupt the silence, but sat spell-bound, always looking at the figure, until I was positively afraid to take my eyes from it.

Again the chimes began their peal for the completion of the last quarter. It was seven o'clock. The moment they ceased, the girl rose from her seat, glanced slowly, sadly, earnestly around, pressed her hands across her eyes, and proceeded in her path toward us. We both stood up as she came near; my friend lifted his hat from his head in the most respectful manner as the maiden passed, while she in return gazed vacantly on him, and, walking slowly by, disappeared in the direction opposite that from which she came. We did not remain, but proceeded with a quickened pace to our lodgings. Arrived there, I asked for an explanation of what we had witnessed.

"Do you remember," asked Partridge, "Alfred Dervilly?"

"Perfectly. He was your room-mate after I left you last winter, and twenty times I have been on the point of inquiring for him, but something at each moment prevented. Where is he?"

"Dead."

"Dead! How? when?"

"Killed by the apparition yonder."

"Nonsense! Do not talk any more in riddles. Out with what you have to say about Dervilly and the apparition, as you call it, and this afternoon's adventure."

"*Bien*, let us light the candles, fasten the doors, close the windows, and take a fresh segar."

This was soon done, and, accommodating himself to his seat in a comfortable manner, my companion commenced the history of

The Fair Mystery.

"Yes—you recollect Dervilly of course, and must remember that before you left us we used to joke him about a fair unknown, who was engaging so much of his time."

"I had forgotten—but I now recall the circumstance; I remember, I was walking with him near the 'Garden,' and he made some trivial excuse to leave me and turn into it. You afterwards told me he had an appointment there, but I thought little of it."

"Well, I will give you the story as I now have it, quite complete, for I was partly in Dervilly's confidence, and was with him during his illness, and when he died. He was born in Louisiana, of French parents, who, after spending some years in America, returned to their native country. He spoke English fluently, as you know, and when you deserted me we became very intimate. Then it was I learned how deeply the poor fellow was in love, actually *in love*. No mere transitory emotion

—no momentary passion for an adventure—no affair of gallantry, was this: his very being was absorbed—he became wholly changed—it seemed as if he had bound himself, body and soul, to some spirit of another world. I never saw, never read, of so engrossing a feeling. At last he confessed to me. He said he had met, a few months before, at the house of a former friend of his family, who had been of considerable consequence under the previous reign, but was now reduced, and lived in obscurity, a creature of most exquisite shape and feature, who proved on acquaintance to be possessed with a loveliness of character, a modesty, an irresistible charm of manner, which took him captive. Dervilly became completely enamoured with Emilie de Coigny. This he discovered to be her name, but on inquiring of the persons at whose house he first met her, he could get no satisfactory information; indeed, a very singular reserve, as poor Dervilly thought, was maintained whenever she was mentioned, so that he could not, in fact, glean the slightest particulars about her. This did not prevent him from confessing his passion, for the girl came frequently to this house, and their acquaintance ripened very fast. Emilie de Coigny felt for the first time that her heart was occupied, and all that restlessness of spirit caused by the unconscious longing of the affections laid at rest, and Alfred Dervilly became the sole object of her hopes, if hopes she had. All this, I repeat, Emilie de Coigny felt; but, singular to say, she hesitated to confess it, even when her lover passionately entreated; it seemed as if something stood between her and happiness, to which she feared to allude. It is not easy to deceive the *heart*, and Dervilly knew, despite the apparent calmness of Emilie, despite her sometimes cold demeanour, that he was loved in return. But one thing troubled and perplexed him; one thing filled him with vague fears and apprehensions, and checked the ecstatic feelings which were ready to overflow within him. A mystery hung about this beautiful girl; she claimed no one for her friend, she spoke of no acquaintances, she never alluded to parents, or to brother or sister, or other relation; she made no mention of her home. Besides, a strange sadness, strange in one so young, seemed to possess her, and to pervade her spirit; and while contemplating that imperturbable countenance, Dervilly at times felt an awe come over him for which he could not account, and which for moments subdued even the force of his passion. It appeared to him then, as if he were under a spell; but presently, when a gentle smile illumined her face, her eyes would be turned on him so lovingly, and her look express, as plainly as look could, that all her trust was in him and in him only. Dervilly would forget every thing in the raptures of such moments; indeed, in his ecstasy he would be driven almost to madness; for of all characters," continued Partridge, "hers was the one to set a youth of ardent temperament absolutely crazy. So matters advanced, or rather, I should say, so time advanced, while affairs did not. It was at this period," said my friend, " that Dervilly gave me his confidence. Our intimacy had gradually increased from the hour of your leaving us, and at length he unbosomed himself completely. My first impression, after hearing his story, was that the pretty mademoiselle was no more nor less than an arrant flirt; that her charms were magnified to a lover's vision; and that the mystery which attended her would turn out to be no mystery at all. So I treated the case lightly, laughed at his description, called Mademoiselle Emilie a coquette, and added, a little seriously, that it was a shame for her to trifle with so warm-hearted a fellow. You know how grating are the disparaging remarks of a friend about one in whom we confess to ourselves a deeper interest than we care to acknowledge to the unsympathizing. What I had said was kindly intended, but it touched Dervilly to the quick.

"'I did not think you capable,' he exclaimed, 'of thus making light of my confidence—I find I was deceived. You are at liberty to make as much sport of me as you will. I have learned a lesson which I will take care to remember.'

"'You must not speak so,' I said; 'I really was not serious. I take back every word. I would not wound you for the world. Forgive me.' Then we shook hands, and Dervilly assured me I had misjudged his Emilie; he would ask her permission to introduce me, and I should see for myself. The permission was never accorded, although Dervilly urged to Mademoiselle de Coigny that I was his best and almost only friend. She was unyielding; she would not see me. Meanwhile his passion increased with every impediment —yet he gained no assurance of its being returned, save what his heart whispered to him.

"In the *Jardin des Plants* they were accustomed to meet daily, when the weather was propitious—so much Emilie yielded to her lover—and spend an hour together; and if they could not meet in the open air, they repaired to the house where they first became acquainted. On one occasion Dervilly, unable to bear suspense any longer, seized her hand, and passionately pledged himself, his existence, his soul, his all, to Emilie de Coigny; he swore his fate was indissolubly linked with hers, that their destiny could not be severed, and he demanded from her an avowal of the truth of what he said. The violence of Dervilly alarmed her; she drew her hand from his, and looking him steadily in the face, inquired:

" 'What has prompted Monsieur to this sudden show of feeling?'

" 'Do you ask what?' exclaimed Dervilly: 'it is *you*. Are you not answered? How can I resist what is inevitable? how curb myself when *all* hold is lost? *Dieu merci!* be not so deadly calm—it means the worst for me—be angry, vexed, any thing, but look not on me with that glazed look—it maddens me.'

" 'Monsieur Dervilly,' said Emilie, without change of tone or manner, 'what you have said, if it means any thing, means every thing; it means all a maiden longs to hear from lips that are beloved. To respond, I must be assured how far your judgment will confirm what now seems to be a mere passionate ebullition. Excuse me,' she continued, as Dervilly made an impatient gesture; 'I have heard and read of similar protestations which had little true significance.'

" 'I accept any conditions,' interrupted the young man, 'and will bless you from the depths of my soul for naming any, even the hardest; yes, the hardest— I care not what, so that they are from you.'

" The girl regarded Dervilly as if she would search his very nature. 'You are silent—speak; I can no longer contain myself,' exclaimed he, wildly.

" 'Monsieur,' once more observed Mademoiselle de Coigny, 'you know not to whom you address yourself;

should I tell you, you would retract all those strong words, and hasten to escape in the least humiliating way possible.'

"'Never. Heaven is my witness, never. I care not who you are; I will never seek to know; when you choose, you shall inform me. You need never tell me. I say, I care not, so that you are mine.'

"'And you will be *mine* for ever?' said the girl slowly.

"'For ever.'

"'I am yours—yours,' and Emilie de Coigny sunk into the arms of her lover.

"In one instant the fortunes of Dervilly were changed: —from despair he was raised to a condition of delicious joy. His raptures were so unnatural, that I cautioned him against such violent indulgence of them. But he was too excited to listen to me. Indeed, I feared that he would lose his reason. It seemed as if more than ordinary passion had possession of him, and that it was inspired by something unearthly; and, without ever having seen the girl, I began to attribute to her a supernatural influence. Besides, Dervilly confessed he knew as little of his affianced as before, and that occasionally the same icy look would be turned on him, as it were quite inadvertently, and hold him spellbound with horror, while it still served to increase his frenzy beyond all bounds. Then, her endearing smiles, her truthful and confiding love, her absolute reliance, her entire dependence, on Dervilly, made him so frantic with happiness, that he lost all capacity to reason.

"The season passed away, but Dervilly had learned nothing more of the history of his betrothed; she still avoided the subject, and, when he alluded to it, she would beg him to desist, and hide her face in his bosom and weep.

"Strange thoughts at last found their way into his brain, fearful surmises began to disturb his peace, and, when absent from Emilie, he would resolve at their next interview, to insist on knowing all. But when the time came, and he met, turned on him, the open and innocent look of the maiden's clear eyes, which expressed so earnestly how entirely her soul rested on his, all courage failed him, and he could not go on. . .

"One evening," continued Partridge, after a pause, and with the tone of a person approaching an unpleasant subject, "one evening, after dinner—I think it was the last week in May—I recollect the day had been quite warm—I strolled into the large garden which you remember belonged to our old lodgings in the *rue*

Copeau, and after a while sat down in the summer-house. Presently little Sophie Lecomte came running out to me, and I remained amusing myself with the child's prattle till it was dark. The moon shone brightly, and I did not perceive how late it was, until reminded of the hour by finding that Sophie was fast asleep in my lap. I rose and carried her into the house, and went quietly to my room. I seated myself near the window without lighting the candles, feeling that the glare would not then harmonize with my feelings. The truth is, I was thinking of you, and of that romantic passage across the Apennines, and of the fair stranger, and so forth. I sat by the window, the moonlight streaming across the room, over the top of the old chapel, the windows and doors open, and every thing still, except the monotonous chirping of a single cricket, louder than that of any French cricket I ever heard before, and which sung the very same song I used to hear when a boy, from under the large kitchen hearthstone at home.

'I began to feel a little lonely, and so started up, and stamped with my feet in order to silence the solitary insect, or arouse the rest of the family; but the old one only sung the harder, and the others would not wake, and I sat down again, and half-closed my eyes in order to lose myself, if I could, in some pleasant revery. My eyes *were* half closed, the perfume from the graperies filled the room, and had a pleasant effect upon my senses, and thus I began to forget where I was and what was about me. Presently I heard a rapid, unsteady step along the corridor; it grew more rapid and more unsteady; I raised my head, and at that instant Dervilly hurried into the room.

"'I knew it—I knew it,' he exclaimed, wildly; 'one of the sirens sent from hell! I have sold myself, body and soul!—I am lost—lost. Ah! I knew it—I knew it.'

"Shocked and surprised as I was by such an extraordinary scene, I did not forget that Dervilly was of a most nervous and excitable temperament. I rose, took hold of him kindly, and asked him what had happened. As I placed my hand on his head, I perceived that the veins were distended, and that the carotid and temporal arteries were throbbing violently. I hastened to strike a light, while he continued to repeat nearly the words I have just mentioned in a wild and incoherent manner. I could now see his countenance, and it seemed as if the destroyer had been ravaging it. His cap was gone. His hair, which was usually so neatly arranged, was tossed over his face in twisted locks; his eyes were fixed, and bloodshot, and sparkling.

"'My dear friend, you are ill—you are excited—let me bring you to your bed;' (we occupied the large room in common, with a small bedroom for each, leading from it;) with this I took his arm, and gently urged him to his apartment.

"'Not there, not there!' he cried, vehemently; "have I not lain *there*, night after night, thinking of her?—have I not dreamed there happy dreams, and seen dear delightful visions? Not there—never—never again!'

"'You shall not,' I said, endeavouring to humour him; 'you shall lie in my bed, and I will watch by you till you are better.'

"The young man burst into tears. This action evidently relieved him, and made him more rational, for he took my arm and I assisted him to bed, and tried to soothe him; but he soon relapsed into an excited fever. Shortly after, he called me to him, and, throwing his arms closely around me, exclaimed, 'Partridge, we were born in the same land; I implore you, by that one common tie, not to leave me an instant; I am a doomed wretch; but save me, save me from the fiend, as long as it is possible.'

"I now became very much alarmed. My first impulse was to administer an opiate; but the case seemed so critical that I determined to send at once for Louis, whose sympathy for the students, you know, is universal. I called to young Stabb, who occupied the next room, and he set off immediately. After a few minutes Dervilly dozed a little; and then he started up, and gazed around, as if attempting to discern some object.

"'Do you wish for any thing?' I said. He took no notice of my question, but continued to glance piercingly in every direction.

"'What do you see?' I asked.

"'*La Morgue!*' he exclaimed, with a shudder, and pointing into the other room—'*la Morgue!*'

"He continued to gaze madly in the same way, still holding his arm outstretched, while his whole frame seemed convulsed with terror; but I could gain no clue to the catastrophe which had fallen so terribly on the ill-fated sufferer.

"It seemed to me an age—it really was but an hour—before Stabb returned. He was accompanied by Louis. You know his skill as a physician, and especially in the treatment of fevers, is world-renowned. I had 'followed' him during the whole of your absence; had become, as a matter of course, one of his warmest admirers; and was fortunate enough to secure his friendship. He also knew Dervilly. Hearing them enter, I stepped into the principal room to meet him.

"'*Mon Dieu! Monsieur Partridge, quel est le mal?*' said Louis, with great feeling. '*Monsieur* Dervilly was at the Hospital in the morning, and I met him as late as six o'clock this afternoon, passing into the *Jardin des Plants.*'

"'God only knows,' I replied. 'Something horrible has suddenly befallen him.' And I gave an account of what had occurred since Dervilly came to his rooms.

"Louis was silent for a moment, and then began to question me very minutely about him, while Stabb went in to keep watch over the poor fellow.—Among other things, I mentioned his love affair; and, believing it to be my duty to do so, I told Louis, briefly, all Dervilly had confided to me. He listened with great attention, and after I had concluded, we passed into the little chamber where Dervilly lay.

"He started up with violence as we came in, as if a severe paroxysm were about to follow. He stared wildly on seeing Louis, and, seizing his hand, he exclaimed, 'Ah, *mon Professeur*, you are a very great man, and you are very kind to come to me, but your knowledge avails nothing here,' touching his forehead. Suddenly he extended his finger, and cried again, '*La Morgue—la Morgue.*'

"'What see you in *la Morgue?*' said Louis, tenderly.

"'See? Her, her!' screamed Dervilly.

"'Who, *mon enfant?*' said the Professor, very gently.

"'Who, but the fiend—the fiend! She has my soul—lost, lost for ever.'

" 'You should not speak so harshly of Mademoiselle de Coigny,' continued Louis, in a soothing tone.

" 'Pronounce not that name: a bait, a trap, a wile of Satan; repeat it, and I will tear you piecemeal!' cried the maniac.

" 'But, *mon pauvre enfant*, what does she at *la Morgue*?'

" '*She?* the fiend—the fiend—sits perched on the top of the wooden rail all night, watching—watching—and when some of the corpses show signs of life, sails down, and sits upon, and strangles them. Keep me away from there. Ah, *mon Professeur*, do not let me go there, to lie on the board, and have her bending over me, eyeing me, watching me, ready to strangle me. *There* again! keep those glazed eyes away—keep them away, I say.'

" All this time Louis was making a minute examination of Dervilly's symptoms.

" The latter presently seemed aware of what he was doing, for he exclaimed, 'The usual symptoms, *eh, mon Professeur?* strongly marked, *n'est ce pas?* Act promptly and decisively, as you say sometimes. Let blood—let blood—*appliquez des sangsues*—ha, ha, ha! that's what we call bleeding, both general and local, ha, ha, ha! then come on with your cold applications: ice, ice, a mountain of ice piled round about the head! follow up with cathartics, refrigerant diaphoretics; after, depleting blister!—say you not so?—blisters to the nape of the neck—blisters behind the ears—shave the scalp—I forgot that—shave the scalp—strange I had not thought of it,—and the hair, *mon Professeur*, I know you will think me very foolish, but——save the hair—I sha'n't have another growth—save the hair. Where was I?—ah, the blisters—that will pretty nearly do for me—keep every thing quiet, very quiet—after a while, digitalis and nitre—digitalis and nitre, *mon Professeur*—have I not said my lesson well?'

"Louis stood perfectly still, regarding the poor fellow with a mournful interest. As Dervilly paused, he took off his spectacles and wiped his eyes. 'Ah, *Monsieur* Louis, you talk very eloquently about medical science, but I baffle you; I am sure of it. Call the class together—*Ah, Notre Dame de Pitie*—call the class together; *voila la clinique.* Thus being thus, it must necessarily be thus. That's a wise saying, *mon Professeur.* Call the class together; propound why of necessity you can do nothing? because of a necessity nothing can be done. Call the class together; be active—vigorously antiphlogistic; time is precious—the patient in danger. Purgatives—I doubt as to purgatives. What think you?' And Dervilly paused, and cast on Louis a look so naturally inquiring, that the latter replied, as it were, involuntarily, '*Moi aussi je doute.*'

"And it was so; with all his genius, all his knowledge, all his experience, and all his skill, the great practitioner stood, while minute after minute was lost, apparently hesitating what to do. At last he called me into the other room. 'Is it not possible to find Mademoiselle de Coigny?' he inquired.

"'I have no means of knowing where to seek her,' I replied. At the same time I remembered she was in the habit of visiting the house in which Dervilly first met her, and fortunately knew the street and number.

"'Let her be sent for instantly,' said Louis. 'Do not go yourself; you may be of service here.' Accordingly I gave Stabb the direction, and instructed him to procure Mademoiselle de Coigny's address, if possible; but if he were unsuccessful in this, to communicate the fact of Dervilly's alarming illness, and beg that Mademoiselle might be immediately summoned.

"We returned to the sick room, and Louis, seating himself in a chair, remained lost in thought for nearly a quarter of an hour, while I did what I could to pacify the sufferer. I could not help wondering that a man, so prompt and so efficient, should lose a moment when the least delay was to be avoided; and as I was reflecting on this, Louis rose so suddenly from his seat that I was startled.

"'There is but one course, and the poor boy has very accurately defined it. Let his head be shaved, and pillowed in ice; bleed him at once—if he faints, all the better.'

"'No danger of that,' shouted Dervilly. 'No syncope with me but the *last* syncope—no syncope—ha, ha, ha! double the ounces—you are timid—no syncope, I say—no syncope.'

"He continued the whole time raving, much in the manner I have described. The room was kept quite dark, and no one was permitted to come in. Louis did not leave the bedside the entire night. Dervilly never slept for an instant.

"On one occasion he started suddenly and threw himself close on one side, and screamed, 'Take her away—take her away!'

"'What is it?' I asked.

"'Do you not see her?' he shrieked, 'sitting on the side of the bed, looking into my eyes; take her away, take her away!'

"I need not detail to you," continued Partridge, "the whole of these fearful scenes. Late in the evening Stabb returned; he had found the house; and although he could not obtain Mademoiselle de Coigny's address, he was promised that his message should be communicated early in the morning.

"'It will be too late,' said Louis, mournfully.

"What a long night it was! The morning dawned at last, but it brought no change to poor Dervilly. I had sent for his nearest relative, who lived over on the Boulevard *Poissonnière*, and was awaiting his arrival with considerable anxiety.

"It was not later than nine. Stabb, the good fellow, had relieved me from my watch, and I was in the sitting-room, in my large arm-chair, still anxious and fearful, when there came a slight tap at the door; it opened, and Emilie de Coigny stood before me. Ah, how beautiful she was, yet how terrified! It was not terror of excitement—mere surface passion—but from the depths of her soul. She was stirred by intense emotion. 'Tell me,' she said, coming earnestly up to me, 'tell me where he is, and what has happened to him!' I put my finger on my lips to prevent her from saying more, and led her to the further corner of the room; but she would not sit down; she begged to be told every thing at once; and I, in a low voice, gave Mademoiselle de Coigny a minute account of all I had witnessed. When I came to Dervilly's exclamation, '*La Morgue—la Morgue*,' the young girl became suddenly very pale, her fortitude forsook her, and she murmured faintly, 'He saw me go in—he saw me go in.'

"I must admit I was, for the moment, not a little tremulous. I recollected stories of devils taking possession of the dead bodies of virgins, in order to lure young men to perdition. I thought of the tale of the German student, who, on retiring with his bride, beheld her head roll from swered the summons at once, and in the most gentle manner endeavoured to persuade Mademoiselle de Coigny to go with her. It was in vain. She would not leave the room. Occasionally, through the day, she would step to Dervilly's bedside, and in the softest, sweetest, gentlest tone I ever heard, say, 'Alfred.' The effect was always the same as at first, exciting the poor fellow to still deeper paroxysms and more violent exclamations.

"On the fourth day he died; the symptoms becoming more and more aggravating, until *coma* supervened to delirium. During the whole period of his sickness Mademoiselle de Coigny never left the house—scarcely the room—Madame Lecomte on two or three occasions almost forcing the wretched girl away to her own apartments. When poor Dervilly sunk into that deep lethargic slumber, so much dreaded by the physician, because so fatal, she came almost joyfully into his chamber, and threw her arms tenderly around him:

"'He sleeps at last,' she said; 'is it not well?'

"I would have given the world for the freedom of bursting into tears, so deeply was I affected by that hopeful, trustful question. What could I do, but shake my head mournfully and hasten out of the place?

"He died, and made no sign; not a word, not a look, not the slightest pressure of the hand, for the one he loved so tenderly, and who watched so anxiously for some slight token. 'Oh,' I exclaimed to myself, as the hardness of such a fate was impressed on me, 'God is just; there is an hereafter; these two *must* meet again.

"Emilie de Coigny left the room where her dead lover lay, only when he himself was borne to his last resting-place. She followed him to the spot where he was buried in *Pere la Chaise*, and remained standing by it after every one else had come away. In this position she was found—standing over the grave—late at night by her friends—some members of the family I have mentioned—who sought her out. She left that splendid city of the dead bereft of reason, and so she has ever since continued. When the day is fine, she invariably keeps her fancied engagement with her lover at the appointed place in the *Jardin des Plants;* she patiently sits the hour, and retires sadly, as you saw her. When the weather is forbidding, she goes to her friend's house and waits the same period, never showing the least symptom of impatience, but, on the contrary, evincing the signs of a bruised but most gentle spirit."

Here Partridge paused, as if at the end of his story.
"Is that all?" said I.
"That is all," he responded.
"Surely not," I continued; "you have said nothing about the strange mystery which killed our poor friend,

and which, as it seems to me, is the main point in the story."

"True enough—it is singular I should have left it out, but it is explained in a word. These same friends of Mademoiselle de Coigny gave me the information.

"It appears that on one inclement night, as the *keeper of the Morgue* was returning from an official visit to the Chief of Police, toward his own quarters, which are adjoining and over the *dead room*, he stumbled over something which a flash of lightning at the instant showed to be the body of a man. He was quite dead, but, nestled down close by his side, with one of her little hands on his face, was a child, about two years of age. Jean Maurice Sorel, although long inured to repulsive sights, had not grown callous to misery. By birth he was considerably above his somewhat ignominious office; he had narrowly escaped with his life when Louis XVI. was brought to the scaffold, for some indiscreet expressions that savoured too much of royalty; yet in the tumults which succeeded, he had, he scarcely knew how, through some influence with the chief of one of the departments, been appointed to this repulsive duty. But, as I have said, his heart was just as kind as ever, after many years discharge of it; and Jean Maurice Sorel, instead of repining at his lot, blessed God daily that he had the means of supporting a wife and children, while so many of his old friends had literally starved to death.

Such was the person who stumbled over the body of the dead man, and discovered the living child beside it. He called at once for assistance, and had the corpse conveyed to his house, while he carried the little girl in his arms. She was too young to give any information about herself, but, on searching the pockets of the deceased, several papers were found which disclosed enough to satisfy Jean Maurice Sorel that in the wasted, attenuated form before him, he beheld his once friend and benefactor the Marquis de Coigny, who, he supposed, had perished by the guillotine in the revolution. The papers permitted no doubt of the fact that the little girl was his grand-daughter and only descendant, and she was commended to the care of the kind-hearted when death should overtake him.

"The old Marquis was buried, and the little Emilie adopted into the family of the good Jean Maurice. Her education was conducted in a manner far superior to that of his own children, and the choicest garments of those which fell to him were selected to be made over for her. Perhaps unwisely, her history was explained to her, so that she lived all her life with the sense that she belonged in a different sphere; not that she was ungrateful or unamiable —quite the contrary—she was sweet-tempered, affectionate, and gentle, and loved by Jean Maurice and all his family with a devoted fondness: but the world had charms for her which the world withheld; she felt that she never could become an object of love where she could love in return, and so she repined at her destiny.

"By accident she made the acquaintance of the family where Dervilly first met her. They had known her father and her grandfather, and she loved them for that. She resisted for a long time the feeling for her lover which she perceived was taking strong hold of her, and, when she could resist no longer, she yet delayed to tell him what a home she inhabited. This was her pride—her weakness—and how terribly did she pay the penalty! Day after day, (so I was told,) she resolved to explain all, but she procrastinated, till her lover, no longer able to restrain his anxiety, and full of excitements and fears and perturbations, followed her at some little distance, just at twilight, and saw, or fancied he saw, her enter the *Morgue*. It was too much for his nervous temperament. His brain caught fire—he came home raving with delirium—and DIED! Now you have the whole."

CHAPTER IX.

CHANGES.

WHEN I came back to Paris,—I alluded in the last chapter to my absence and return,—I found most of our old company still there, but occupying other quarters. In justice to our friend, *Monsieur* Battz, and his interesting daughters, I should say that it was through no fault or inattention of theirs, but from the mere desire of change, that our clique one day made their exodus from the *rue Copeau*, and took possession of a habitation in an adjoining street.

To be sure, there were some advantages in the new location. The house—it was a very large one—had been unoccupied for nearly two years: it had the reputation of being haunted, as a matter of course. The young fellow who kept the billiard-room, where some of our party used to congregate, had married, and at a venture rented the old cobweb-covered mansion. By the change we lost the cockney, and two or three others; a few also of our companions had left Paris, and their places had been supplied by others. By the time I got back every thing was settled, except the ghosts: according to the Italian, who occupied with his friend a remote part of the building, and who delighted always in the marvellous, there were strange doings every night, immediately after twelve o'clock, in the long corridor which ran by his room. No one could tell whether the Italian was serious or jesting; he was the first to pass a night in the house, and claimed to have more authentic information than the rest about this very delicate subject.

For myself, I recommenced my walks, and became again very regular in my pursuits, so that even Partridge was fain to commend me. Next to him, none were so well pleased to welcome me as Clements; he had a great deal to tell me, and our intimacy became stronger than ever. One evening several of us happened to meet in the Italian's room: the latter appeared for a time in much better spirits than usual, and amused us with many laughable reminiscences of his life, and what he had seen in different countries.

"What has put you in such good humour to-night, *Signor Italiano?*" asked one.

"Nothing but a good dinner and a good digestion," answered another. "It always affects the Signor wonderfully."

"Right, quite right," said the Italian; "we have a proverb which I have seen also in the English: 'An overloaded stomach talks this planet into hell—a glass of wine can deify its devils.'"

"Ah, now we understand the ghost stories," said Clements.

"*Messieurs*," said the Italian, "you get no ghost story out of me. You are a set of unbelievers; I shall not give you a fresh opportunity to scoff."

"Come, Clements, let us be off; we shall make nothing here to-night."

"Stop a moment," said Vincent, who had just come in, "I have a letter from Howard, from whom we have not heard for an age. He is our melancholy Jacques, you know; that is, on occasions when Howard affects the character, because he thinks he writes well, and that it makes him in-ter-est-ing. Here is the letter, written

'Under the shade of melancholy boughs

It gives a pithy account of the New York atmosphere. What a fool he is for a fellow of sense! Hear him:" and Vincent read from the letter:

"'You have no idea, my friend, of the insensate follies of a New York "season." The people are wild without being gay, and excited without being animated. They are extravagant without taste, and profuse without generosity. Imagine every thing that is *un*natural, and you shall not fail to get an idea of "society" in my native city.'"

"The fellow is in love with some pretty New Yorker," said Partridge.

"Don't interrupt," cried Vincent; "listen to what's coming. It's poetry, by Jupiter Ammon:

> '"There they are happiest, who dissemble best
> Their weariness; and they the most polite,
> Who squander time and treasure with a smile,
> Though at their own destruction. She that asks
> Her dear five hundred friends, contemns them all,
> And hates their coming. They (what can they less?)
> Make just reprisals; and with cringe and shrug,
> And bow obsequious, hide their hate of her."

Week after week we had in our own set one continual scene of bustle and bewilderment. Parties succeeded parties; and dinners and suppers and dances made up the rest. Four brides caused all this tumult; more, probably, than they ever can make again: unless, in the mysterious course of events, they should all die the same season, and within the same fortnight, and in the same city. Even then, methinks, the stir and noise would be nothing to what it has been. The friends doubtless would "sympathize," and perhaps some startled youth might mutter to himself as he passed along—

"This looks not like a nuptial;"

but the current of oblivion would flow smoothly on and over them; the summer's grass would grow green upon their graves, and the winter's snow heap forgetfulness upon their turf.—How are the Battzes? Where's Milor *Anglais?* Who knows any thing about any body? Send me the end of the segar that Alibaud left unsmoked, just as he was submitting to be guillotined; I want to preserve it in a glass case. Do you know, that affair always puts me in mind of Barnardine: "Master Barnardine," says the clown, "you must rise and be hanged, Master Barnardine!" Whereto he replies, "Away, you rogue, away; I am sleepy!" but the clown persists, "Pray, Master Barnardine, awake till you are executed, and sleep afterwards." Don't forget the stump of that segar. N. B. What about the compound fracture? did the lad recover? If he didn't, C—— killed him; I say he *killed* him. How is the roll-call? Remember me to the boys each and every. Adieu.'

"Now," said Vincent, "were it not for his unbearable affectation——but he is absent; we won't make his ears burn. Let's drink his health."

"And then leave me alone with the 'genius of the house,' I suppose," said the Italian.

"Alone? there is your friend!" pointing to the Genoese, who was asleep on the couch.

"Slumber is a temporary death: I am worse than alone in such a case."

"Oh, aye! but the 'genius of the house;' what of *her?*

We are to have a description one of these evenings, I suppose?"

"No, you are not; at least, not till you will approach the subject with more reverence. But medical students and medical men are a set of materialists—a miserable set too. I pity the whole race; and particularly because they are expected to do so much, and can really do so little. Voltaire, carrying out this idea, pronounced a physician to be an unfortunate gentleman who is called every day to perform a miracle—'reconcile health with intemperance.' He was more charitable than Talleyrand, who always declined to recommend a cook or a medical man, because he did not wish to be held guilty of murder as an accessary before the fact!"

"Hallo! what is the matter with the *Signor?* Allow me to feel your pulse!" and Vincent drew out his watch with a professional air, and commenced counting.

"*Signor Italiano,* you are very sick indeed; judging from present appearances, I should say your life might reasonably be despaired of."

"Have done with your nonsense," said the Italian. "You won't have me to practise it on much longer, however. We are off!"

"Off! how is that?" cried several.

"I am tired waiting for revolution in Europe: we are going to a land of freemen—to your country, Mr. Vincent

—the United States of North America. I have looked at the signs of the times; it is of no use for us to wait. I was here in '30. Blood was not poured out in vain then. It was but the first step; since then it has been the half step backward. But it is coming—it *is* coming! We shall be recalled from America years hence to fight the battle of freedom—perhaps in these very streets. Who knows?"

The Italian paused; his fine countenance was lighted by a generous fire; his eyes were steadily fixed in the distance, as if attempting to penetrate the future. I could not but say to myself, that there yet burned some of the spirit of ancient Rome in the breasts of those whom we are apt to call her degenerate sons.

"Yes," continued the Italian, "we go to America. As pilgrims seek a shrine, so seek we. Once there, we shall breathe again with a sense of freedom, while the thought of home, when the waning day seeks repose in the Occident, will fill our hearts with a gentle sadness, instead of the bitterness we now feel.

> ' Era già l'ora che volge 'l disio,
> A' naviganti e'ntenerisce il cuore,
> Lo dì ch' han detto a' dolci amici addio,
> E che lo nuovo peregrin d'amore
> Punge, se ode squilla di lontano,
> Che paja 'l giorno pianger che si muore.' "

The words of the Italian sensibly affected the whole party. I have before mentioned that no one appeared to know precisely about him or his companion; both were considerably older than any of us. As Louis Philippe was at that time very tolerant of refugees, Paris contained an unusual number of them; and no one thought it best to ask questions.

"*Signor*," said Von Herberg, after a few minutes, in which we were all silent, "have you ever come to any different conclusion about what you beheld on the Boulevards one night of the Revolution?"

"I still hold to the very same; my opinion has not changed in the slightest. The day *will* come."

"What is it?" whispered one.

"I do not know," said another.

"What were you speaking of, Von Herberg?"

"He speaks, *Messieurs*," said the Italian, emphatically, "of what I beheld on the evening of the last of the 'Three Days.'"

"What? what?"

"You know at that time there was some hard fighting: the trees on many of the Boulevards were cut down, and barricades were made of them, with the aid of coaches and omnibuses, and other carriages. It happened frequently that the people had not time to carry away their dead; so they would deposit the bodies occasionally in a position that they might neither be trampled on, nor passed by unnoticed when occasion should permit their being removed. I was going along the Boulevard *du Temple* the evening in question. At the principal barricade an immense tree with large branches lay stretched entirely across the side-walk. As I endeavoured to work my way through, I encountered a man planted bolt upright against one of the limbs of the tree; a lantern was burning near, and cast its light across his features: a second glance discovered to me that he was dead. I had seen similar sights, and this did not startle me. I proceeded on my way. In half-an-hour I came back, and passed the same spot. There were *two* men placed where I saw the one; each was the exact counterpart of the other, in every particular; just alike— exactly alike! I halted so near that I could touch them. I shut my eyes and opened them again; it made no difference. I pinched myself, to be certain I was not in a trance; I soon satisfied myself on that point. Then I rubbed my eyes very briskly; still there stood the *two!* It was then, after every other trial had failed, that I put out my hand to *touch* the bodies. I extended it to the one nearest me, when suddenly the other raised its arm, and, with a menacing gesture, interposed its hand between me and the dead man. I was perfectly calm, *Messieurs*, because I felt conscious of no ill: I deliberately dropped my hand, and at once the arm of the other assumed its original place.

I was determined to probe the matter. I stepped a little nearer, to take hold of the body which had made such a strange demonstration. I extended my hand so that it would rest on the shoulder of the other: it encountered *nothing;* but fell by its own weight quite heavily to my side. Still the *appearance* remained; and after another look, I disengaged myself from the branches, and came away."

"A very interesting case of optical illusion," said one.

"Very," responded another.

"Yes, indeed!" exclaimed a third.

"*Messieurs,*" said the Italian, warmly, "there was no illusion about it: I was as cool and as collected as I now am. I tell you I beheld the *anima* of the dead citizen. It was an omen that our cause—the sacred cause of Freedom—LIVED! and so I hailed it; and so I still hail it! The day WILL come!"

"Ah! well," cried Vincent, "I don't pretend to judge of these things. Somehow, those who want to see ghosts, always can see ghosts; and those who are unbelievers, as you say, are not troubled with them. For myself, I prefer not to be troubled. But what a break-up we shall have! I go to New York next month. Clements, you are going to London?"

"Yes, and shall take Partridge with me to 'walk' Guy's."

"Let us see," continued Vincent; "*Signor Italiano* and the Genoese off too! By the by, we must manage to go together. And two left yesterday: it will be a regular clearing out!"

Von Herberg and I looked at each other. "We must stick together," I said.

"We will!"

In another month our whole society were scattered.

CHAPTER X.

NEW QUARTERS.

The scene has changed. Franz von Herberg and myself occupy pleasant apartments in the *rue de la Chaussée d'Antin;* quite *tout en haut,* to be sure, in order to give Franz a better arrangement for his canvass: yet the situation is for the time certainly a delightful one. Partridge and Clements are in London. The former is determined to compare practically the English and French methods of treatment. He writes me he is charmed with Astley Cooper, and that he likes Key. Clements is not satisfied quite. No Englishman ever admits that he is entirely pleased in his own country, and out of it, *every thing* is wrong. I do not mean this as applicable to my friend, for he is essentially a cosmopolite.

I have become very much attached to Franz. He is a congenial companion; a true artist; and what is more, he is a German without being mystical. We are almost inseparable.

A narrow balcony runs before our windows, just wide

enough to admit a chair: here we sit and converse, or watch what is passing;

"And dizzy 'tis, to cast one's eyes so low!"

Sometimes I direct my attention to our neighbours opposite. Those directly in front are a comfortable-looking old couple, without "chick or child:" they spend nearly the entire day playing backgammon. They are playing in the morning as I take a look across after breakfast: they play during the day incessantly. The old gentleman goes out about twelve; he returns in two hours, and they commence playing again. After dinner both go out together; and when they come in they begin once more. So they have gone on for weeks. It makes me nervous. I have a restless, unconquerable desire to rush over, seize board and dice and boxes, and toss them out of the window. Why *won't* they stop playing? Can such a sight be witnessed any where but in Paris?

The rooms next to the backgammon players are occupied by two nice-looking grisettes. How much taste is displayed in the arrangement of their simple furniture! Outside, on the ridge formed by the retreating roof, are displayed a row of flower-pots: I was about to say the plants are cultivated with great care, but nothing like *care* is manifested. They are looked after and cherished with the same tenderness one would wait upon some living thing.

These girls are evidently sisters. They rise early, and before breakfast they come to their flowers,

> "To visit how they prosper, bud, and bloom."

They talk to them—they caress them—they watch every bud; they mourn if some noxious insect has, unperceived, committed any depredations. Occasionally a new plant is brought home, and then such an excitement is produced! I can easily imagine that these flowers grow the gladlier under such "fair tendance." After breakfast they put on their neat little caps, and go to their labours: they work all day, and come back at night as cheerful as crickets.

On the other side of our players lives an old lady with an idiot son. He is grown up. He seems quite harmless. The poor woman is very devoted to him. In the morning she attends to his toilet, washes his face, combs his hair, and places his chair for him. Then she prepares his breakfast, and feeds him as she would an infant. He never shows any emotion, except to betray his satisfaction by a hideous grin, and his dislike by strange, unearthly exclamations. His mother loves him—loves this abortion! She caresses him: I see her do so daily. Yes, that idiot is *loved*. He can return no affection: he can feel none.

Poor lad! Poor woman! Why do I say "poor lad!" "poor woman!" What right have I to say so? God only knows whether it be so or not. God help them, and forgive me!

One "flat" lower down, and I see a comfortable family who belong to the shopkeeping class, all of whom are turning their hands to something. How gayly they sally forth Sunday morning to mass; and in the afternoon for an excursion in the gardens, or perhaps a little way out of town.

Lower still, if I count correctly, *au troisième*, I perceive very fine people—fashionable people—with exquisite furniture, mirrors, curtains, paintings. They live, one would suppose, expensively; and yet every *sous* is calculated as closely and as systematically here as by their neighbours *tout en haut*. Strange as it may seem, notwithstanding the elegance of the repast, which is daily served at five o'clock, I would lay an even wager that the unexpected presence of two friends at the dinner-table would endanger the sufficiency of the supply, and put the family to inconvenience. From high to low the French are the most economical people on the face of the earth. But this is not *romance*.

"Franz," said I, one morning, as we were returning

from the inspection of one of David's paintings in a private collection which my companion desired me to see, "Franz, you recollect you were trying to paint something, I do not know what, when we were in the *rue Copeau*, which you then found it impossible to finish. I have wanted very often to ask you what it was, and whether you have since completed it; pray tell me now."

"Simply this," said Von Herberg. "I was at one time in the habit of attending service at the church *Notre Dame de Lorette*. I was first attracted there by the music, and afterwards by the eloquence of a young man, who was the only priest in Paris that I ever listened to with interest. One day, as the people were moving out of church, I saw a commotion near one of the side-chapels. I went to the spot. An old mendicant, who had for a long time been in the habit of frequenting the place, had just been discovered, leaning against the wall, in a kneeling posture, but quite in a lifeless state.

"It seemed as if vitality were lingering about him when I came up, for there remained on his features a certain *living* expression, worn doubtless during the last moments of existence. I cannot describe it to you. There was nothing repulsive—nothing disagreeable in it; but such as you would imagine a weary wretch to exhibit when about to be freed from the load of life, and transported into those regions of bliss which faith has made clear to him. Ah! if I could only depict that! It was in vain. I tried, and tried again, but could do nothing with it. By the way, do you not believe some agency might be introduced to bring back the escaped or escaping spirit? May we not look for some wonders yet through the aids of electricity?"

I shook my head.

"Why not?" continued Von Herberg. "Life has been compared to a candle. Now I cannot better illustrate my meaning than by referring to it. Extinguish a candle, and you easily relight it, without any direct contact, by applying a torch to the column of smoke which rises from it, even at a considerable distance. So it has seemed to me that vitality might, by electrical process, be brought back, if application should be made seasonably. And such appeared to be the situation of the beggar in the church of our Lady of Lorette when I first beheld him.—Strange that I could not catch that expression!"

"There are many reasons," I replied, "why the analogy should fail; although I confess I am struck by the way you present it: but after all, disguise it as you will, it is no more nor less than rank materialism. I abominate it! I shudder at it! No man hath power to *retain* the spirit, much less reclaim it. Indeed, very apropos of this are the lines of Sir Richard Blackmore; (Von Herberg understood English well;) let me repeat them:

> ' A flowing river, or a standing lake,
> May their dry banks and naked shores forsake ;
> Their waters may exhale and upward move,
> Their channel leave to roll in clouds above;
> But the returning winter will restore
> What in the summer they had lost before:
> But if, O man, thy vital streams desert
> Their purple channels, and defraud the heart,
> With fresh recruits they ne'er will be supplied,
> Nor feel their leaping life's returning tide.' "

"Those are very fine," said my friend; "but I do not think they are directly applicable. Perhaps they are though. What a mystery is this *dying*. How on a sudden was our beggar promoted over all who surrounded him. What notice he attracted, too, for once! Not a soul would have turned their head around for him while alive, yet how they all thronged around with their 'sympathies' when he was dead. How disappointed might some have been if he had revived, and made personal application for relief. But here we are at home. Where shall we dine to-day?"

"What say you to Champaux?"

"So be it: let us go.—I shall never get that out of my brain till I can get it on canvass."

"Perhaps you will be more successful now that I have roused you into a new excitement."

"It may be so, but I do not want to be excited."

"Dinner will prove a sedative."

"So I hope. Come."

CHAPTER XI.

THE CAFÉ.

We were seated, leisurely discussing the merits of Champaux's *carte*, when I heard a loud voice near us, which attracted our attention so much that we turned to listen to it.

"Garsong, why the deuce don't you *venez ici?*"

The waiter came up.

"Do you suppose I am going to *manger* such dishwater stuff? What do—a—a—*s'appeller cela?*"

"*Potage, Monsieur.*"

"Pottage! Now do you a—a—*comprenez?* I don't want pottage—I want soup! Do you hear that—a—a—*entendez vous?*"

"*Oui, Monsieur.*"

"Then why the devil don't you make tracks for it, eh?"

The waiter stood in mute astonishment, with a permanent shrug on his left shoulder.

"I say," continued the other, "what are you standing there for? Where are the pizés?"

"*Des pois, Monsieur?*" said the poor *garçon*, catching at the word; "*oui, Monsieur;*" and he was hurrying off.

"Stop! a—a—*arretez!*" said our character, catching the other by the arm; "what are you after now?"

The *garçon* cast an expression of mute despair over the room. Happening to catch our eyes, (for I must say we were enjoying the scene immensely,) he assumed such an appealing look that I rose and stepped forward to act as interpreter, when all at once I recognised in the individual a good-natured, rattle-brained, go-ahead New Yorker, to whom I was introduced a few weeks previous, and who had come out on business to England, and was determined, as he said, to have his own fun, and see Paris, if he didn't know the language. He greeted me immediately. The usual congratulatory expressions passed, and I hastened to introduce Wilcox to Von Herberg, and transferred him without ceremony to our table. After that, I inquired how he had been since I last saw him?

"How have I *been?* I have been starving—slowly, gradually starving to *death!* Look at me!" and Wilcox put his hand over his large, fat face, and across his stout arms. "Yes; ever since I have come to this infernal place, I have been trying to get one substantial *meal of victuals;* and I tell you I CAN'T DO IT!"

Here the *garçon*, who had taken the opportunity to absent himself as soon as he saw us engage in conversation, returned with the plate of peas which our friend had ordered.

"There!" exclaimed Wilcox, "do you see that? This is what they call a plate of peas—plate pizés, I suppose I should say. Now look at them. Do you see"—taking up a large table-spoon—"I can put the whole 'plate' in this spoon and swallow them at one mouthful. Here, garsong, bring me a plate pizés, American—large—*gros, comme ça*. By George, I am getting desperate. I want something to EAT! And there's something else I want; I want a bottle of Scotch ale. I would give this minute a guinea for a bottle of Scotch ale—a good, stiff, quart bottle of Scotch ale. *Can* it be got in this city?"

"Yes; I will give you the direction where you can have the genuine article."

"Then I am off!"—seizing his hat—" but stop; now you are here I will make one more effort for something to eat. No Scotch here, I suppose?"

I shook my head, while Von Herberg suggested that he could order a bottle of beer.

"No you don't!" exclaimed Wilcox. "I want none of that wishy-washy stuff. I thought yesterday I had found something which would go to the right spot. I called on the ale—the boy brought me a great big bottle, *comme ça,* (lifting up his hands,) which held about two quarts. I began to lick my lips over it. The cork was drawn—my tumbler filled. I was thirsty; understand that. I fixed my eyes on the garsong, and I began to drink. The dog looked guilty, and was about to sneak away. I gulped two swallows before I knew what I was doing. I set down the glass. 'Boy!' I shouted, for I was too much excited to speak French—'boy! what's this you have been giving me?' And what do you think it was?" said Wilcox—"for the poor devil was too frightened to answer me—what do you think it was?"

"Beer, I suppose."

"*Beer?* I should call it a compound of water and molasses kept just long enough to be a little sour. How much do you suppose they charged me for it—two quarts at least? I will tell you—*ten cents*, ha, ha, ha! ten cents, as I am a live man, ha, ha, ha! I put up the money, and sloped—glad to get off so. But what are you eating, eh? I see—a mutton-chop. Speaking of eating, some lads I fell in with here said they would let me into the secret of dining well, and a fanciful way of getting a dinner it is. The party that came over with me all manage it that fashion, and it's after this style. The plan is for five to go together and order dinner for *three*. In this way they say they get a variety. Egad, I am thinking I could better the system—let every man go by himself and order for *five*. That *is* a good plan, and it has just struck me—I'll carry it out. Garsong! you little vagabond, come here— a—a—*desservir*—a—a—curse the pottage—off with it; that's plain American. Now, let us see: roast beef—no go; beef-steak—can't cook it; mutton-chops—first rate, if I could only have enough of them. Just tell this fellow to bring mutton-chops for five and potatoes to match."

I looked incredulous. "Upon my word I mean it. I pledge you my honour I am dying from HUNGER! Tell him to be in a hurry."

As Wilcox was obstinately set on having his way, I gave the order with an explanation, to the *garçon*, that our friend was a mad wag who wanted to indulge in his joke. The *potage* was removed, and the chops and potatoes actually served, and, what is more, were *eaten*. Badinage apart, I really believe that Wilcox was not only hungry, but that he had really suffered from the manner he had been treated to French dishes.

"Have you been in Paris the whole time since I met you?"

"No," said Wilcox, emphatically; "and that's what I want to tell you about—I am going to make a grand business of it. I undertook to go to the south of France, for I didn't care about being home before cold weather, and as I was improving so much in French, I thought I would venture it. I got on well enough to Lyons, for there was an acquaintance of mine going there, who knew the country well.

"The morning after I reached Lyons, I started for Marseilles, when, about half way, we came to a small, dirty town, with a narrow stone gateway for an entrance. I cannot remember the name of the place; I don't want to remember it; indeed, I don't believe I ever knew. Well, we halted at the gate. Our passports were called for and taken from us as usual, and we cracked into a little tavern and stopped. I thought at the time one of the guards eyed me suspiciously. Presently a soldier came up to me. He could speak English a little.

"'*Monsieur* is an Englishman,' he said.

"I shook my head. 'American,' says I.

"At that he shrugged his shoulders and said, '*Monsieur* cannot proceed.'

"'Why not?'

"'Passport has not the *visé* of the Prefecture of Police at Paris. *Monsieur* must remain here.'

"A pretty muss I was in, to be sure; but that was not the worst of it. It turned out that an Englishman had, a little before, left Paris, who was accused of a treasonable correspondence with some of the cursed factions opposed to the government, and it became important to arrest him. A description of his person had been sent all over the country, and, what was deucedly unlucky, it answered almost precisely to me. Of course I protested in the most vigorous terms that mortal man could invent. If you understand any thing about a Frenchman, you should know that the more importunate *you* are, the more dogged is *he*. The more excited you become, the more indifferent he grows. I could not move the rascal. He referred me to the mayor; and, guarded like a felon, I was introduced to that dignitary. Of course he was on the alert for a criminal; and, once more of course, I *was* the criminal. I argued, I entreated, I explained, I insisted. It was of no use. 'American citizen' had no terror for *Monsieur le Maire.* I wanted to send to our Consul at Marseilles. The ignorant, stubborn old fool said it was altogether unnecessary. It was a very simple business. My passport was to go back to Paris, and I was to go to the town jail, or whatever you call it. If the Prefecture of Police said 'all right,' and affixed his *visé,* then all right it would be; otherwise, I was certainly the 'Englishman,' no matter what the American Consul said on the subject.

"'But,' urged I, endeavouring to keep my temper, 'suppose my passport should happen to come back "all right," what excuse would you have for treating me in this outrageous manner?'

"I was answered by a shrug, and a cursed impudent, incredulous gesture, but not a word would the old devil *say*. I tried him again and again; he grew worse and worse, until he ceased to notice me at all.

"The result was, I was marched off to the jail— a most dirty old building, with a heavy stone archway, over which were inscribed certain words which I sha'n't soon forget. I am a pretty good French scholar—you needn't smile—and it didn't take me long to read them, and I believe the malicious puppies halted on purpose so that I should. I said I never should forget the words. No more shall I; but, for fear I might, I took occasion to put them down among my memorandums." Wilcox pulled out of his pocket a small note-book, opened it, and, putting his finger upon a line he had pencilled, said, "There you have it."

We both read aloud in the same breath,

"*Ici on se repent, mais il est trop tard;*"

And both of us burst into a laugh, despite the wanton lack of sympathy which it manifested.

"Gentlemen," continued Wilcox, "it was no laughing matter, let me tell you that. After I had spelled it out, my teeth began to chatter, for I did not know what these cannibals were going to do with me. Well, I was marched into a narrow hall, from each side of which doors opened upon loathsome cells, and into one of these I was thrust. I believe I began some hideous lamentations, for a horrible-looking wretch approached me from one corner of the cell—he was my messmate, you understand—and in very tolerable English endeavoured to console me. 'A cove must expect to be lodged once in a while; I must put a good face on the business. Keep quiet, take it easy, never say die—it might have been worse.'

"I sat down on the miserable boards where I was to lie, and which were covered with a single blanket, and undertook to explain to the fellow that I was no criminal, but was most unjustly and unwarrantably incarcerated.

"'Oh, certainly, of course; but you need not be afraid of me—I never peaches.'

"At that instant I started to my feet as if I were shot, and gave a bound that sent my head against the top of the cell; then I commenced pulling at my clothes.

"'What is the matter with the poor boy?' said my vagabond, in a comforting way. 'It's nothing but the fleas, do you see. After you have been here a while you'll get used to them.'

"I yelled with vexation: I wanted to beat my head to pieces against the door, but at last I flung myself in despair on the loathsome bench, and gave myself up soul and body to the fleas : as somebody once said, if they had been unanimous, they would have lifted me out of bed. I expected to die there; I sometimes think I *did* die, and have not come to life again.

" On the third day I got a request forwarded to the mayor demanding an interview: much to my surprise he came to me, was tolerably civil, but perfectly unmoveable on the subject of setting me at liberty. He had sent my passport to Paris, and in due time the case would be attended to, and the old villain made me a low bow and took himself off.

" On the same day my vagabond messmate was set at liberty, and I prepared a short note addressed to Americans, Englishmen, and to the American Consul, which the scamp promised to deliver to some proper person, if he had to walk all the way to Marseilles to do it.

" After he left I was in better spirits—judge of my astonishment, however, when at night Monsieur Tonson was brought back, having been caught in attempting to commit some petty theft the moment he turned his back on the town. He was searched, and my letter was found on him, and as it spoke of *Monsieur le Maire* as an unmitigated ass, villain, fool, scoundrel, and what not, I expected to be summarily dealt with. Here I was mistaken—the old donkey came to the jail, brought me my letter, and, without a word of comment, marched off. Luckily, he was too firmly intrenched in his own conceit to be moved by it.

"But I was in luck after all. A rumour of the matter found its way among the passengers of the next diligence. There was *one* genuine Yankee among them. —'I'll stand by that chap as long as *I* live'—he could speak French like a native—he insisted on visiting me. I gave him the whole story. He went to the magistrate, declared to him that he knew me, and all that sort of thing, and demanded my liberty. Although this shook the old fellow's faith in my being the Englishman, he would not liberate me, but I got a better room forthwith, and was treated with some decency.

"My friend hurried on to Paris, had the matter overhauled forthwith—it would have taken a month as it was going on—and in three or four days more I was released. Now, what do you think I did? I had vowed vengeance on the mayor, and determined to take the first instalment out in heavy curses over his head and shoulders, well laid on. But the old fellow came to me, and in measured terms tendered his regret at what had occurred, as if it was entirely a matter of necessity, and at the same time asked me to dine with him, with such profound gravity, that I was completely upset. I couldn't stand the dining. I declined—but how could I swear at him after that?

"I took the next diligence for Paris, and have since my return been, as I told you, fairly wasting away under the effects of starvation. I have eaten something now, I'll go and get the ale, and I am thinking tomorrow I will *vamose*. Call on me, will you, when you come to New York?"

We separated. I have never seen this curious fellow since. He went, I understood, shortly after to South America, and that's the last I ever heard of him.

As Von Herberg and I were returning from the *café* to our lodgings, we saw preparations for a funeral before one of the finest houses. My friend took my arm, and we stepped up the staircase and into the room where the dead lay. There was already a good many in the apartment. The coffin was placed in the centre, and immense wax candles were ranged around it, throwing their rays over the darkened room. The splendid mirrors were covered, so as not to reflect the countenance of the deceased, and so shock those present. Magnificent bouquets, purchased at a large expense, were laid on the richly ornamented coffin, but I saw no simple flowers strewed over it; indeed, every thing had reference to ostentation and display. Not one of the proprieties of a funeral were omitted; and, that the departed might be properly assoiled, an unusual number of priests were in attendance. Were it not that the man was *dead*, the spectacle would have been rather an agreeable one than otherwise. Those in the room were becomingly *triste*, while nobody seemed to mourn.

We contemplated the scene for a while, then descended to the street again.

"What think you of this idea of endeavouring to present death in a less formidable shape?"

"I don't believe in it," answered my friend. "Death is a *terrible* event, and it ought to be so regarded by us. Any attempt to dilute the effect produced on us by the great Destroyer seems to me unnatural in the extreme. Every thing here is overwrought. Affection may dictate the planting of a flower on the grave of those we love, or scattering fresh-gathered blossoms over it; but when shops are erected to *manufacture* these tokens of remembrance, when one *pays* for the gathering and the arranging of the garlands,—nay, when the very flowers of which they are composed are artificial,—I consider it a sacrilegious mockery of real grief, and of the feelings of the sincere mourner."

"I think so."

"Strange," continued Von Herberg, "that even at the last moment—in death itself—every thing is done that *can* be done to relieve against the indignity of dying. Gorgeous funerals, costly grave-clothes, magnificent monuments; in every thing the artificial for the natural. What was first a mark of real affliction has come to take its place altogether. A righteous retribution, when we attempt to give form and substance to feelings which are at once destroyed by exposure and parade."

The subject was not a cheerful one, and I did not encourage Von Herberg to pursue it; for he was always too much inclined to fall into a melancholy mood.

We had wandered in the direction of the *Tuilleries*, and were brought back to pleasing visions of this world by the clear, merry laughter of the children, who were running and skipping from place to place in all the exuberance of young life.

CHAPTER XII.

ALMOST AT THE END.

STRANGE to say, we soon tired of the fashionable part of Paris, and had we purposed to remain for a much longer period, I do think we should have sought our old quarters. As it was, after spending a few weeks in looking at all that was worthy of observation, as well in the streets as out of them, we undertook several short excursions into the surrounding country, sufficiently far from Paris to be out of the reach of its immediate influence. These excursions we enjoyed exceedingly; I will, however, give an account of but one of them. As it is impossible to prejudge the effect of a work upon the reader, I have thought it would be judicious to bring my volume to a close before it reached a length which should make it particularly ponderous if it met with disfavour; while, on the other hand, should it prove acceptable to any, I shall take leave of such while the impression is still an agreeable one.

We Prepare to Leave Paris.

The autumn had come round again. Partridge, having finished his prescribed course in London, now rejoined us. We were to spend the winter in Germany. In the spring Partridge was to return to America, and locate in Philadelphia.—Why may I not, even here, pay a passing tribute to his subsequent career. Thank God, he still lives, enjoying, as a practising physician, the reward of his patient, scrutinizing investigations in almost every hospital in Europe—an old and long-tried and attached friend.

We were preparing for our departure. While Partridge was out attending to some commissions, Von Herberg brought into my room a picture, which he had just finished. He had purposely kept it out of sight till it was completed, and now it stood on the table perfect—absolutely perfect.

At this moment Partridge came in. He was attracted at once by the painting. He wanted to know all about it. There was some incident connected with it—he knew there was.

I was, however, in no haste to explain. I remembered the summary way I was dragged from Calais when I was so desirous of loitering on the road, so I took the opportunity of teasing my friend for a few minutes before satisfying his curiosity.

There was nothing peculiar about it—quite a fancy-piece.

"No such thing."

"But, my dear fellow," I continued, "why are you so particularly curious about this little painting? I do not see any thing to justify your stubborn assertion that it is *not* a fancy-sketch, and that there *must* be some story connected with it. With what shrewd appreciation you take in the whole group! A vine-growing country, for the vineyards extend in every direction, almost surrounding the old chapel, over whose entrance is carved in wood an image of the patron saint. The doors are open, and around them still linger two or three old people and a few children, while the solitary figure on this side, you maintain, bears a positive unmistakable likeness to me! How ridiculous this idea of yours: really, you are carrying your discrimination quite too far. You will not give it up? Ah, the picture again diverts you. The foreground embraces a gay company; evidently a wedding-party; rustical, to be sure, but so much the more charming. How *very* joyous seems the occasion——"

"'Fancy-piece'—nonsense!" interrupted Partridge. "Look at the bride: no painter nowadays could limn that face and form from his imagination, nor pourtray the blissful satisfaction which beams in the manly countenance of the groom. There is an evident truthfulness in the grouping, in the portraits, in the expression of each person, in the careful attention paid to details, which cannot deceive me. The story—the story—let us have that!"

"Positively there is none. Still, if affairs had taken another turn—thank God they did not; but as it is, there is no tale of blasted hopes nor of. broken hearts, nor of untimely sorrows which are only quenched in death. Nothing of these. You are still not satisfied? You ask for the merest explanation of the scene; that will content you. Well, perhaps for once I may act the hero; you have all told your stories; why may not I tell mine? Sit down, my dear boy—have patience, Franz. *Silence!* I am going to begin.

The Story of Marie Laforet.

AMONG the numerous *Passages* which so frequently connect one street with another in the finer parts of Paris, and which, as you know, are adorned on each side with exquisite little shops, containing every thing in the way of vendibles that can be made attractive, the *Passage des Panoramas*, Franz and I being judges, is the one most attractive, as well as most frequented, not only by the fashionables of the city, but by the strangers who congregate here. The significant words, "*English spoken,*" placarded here and there, draw to the spot many of the subjects of *la perfide Albion*, who, while they will not condescend to learn the "miserable language," can scarcely do without French gloves and French shoes, and, I might add, French every thing. Now, my friend, I do not affirm that the English tongue is spoken in all its purity at the places where this magic sign is exhibited; on the contrary, I am sorry to bear witness that often it is positively a false pretence, where, for example, the speaking of English is confined to "What you want, sir?" which being delivered, the pretty grisette trusts entirely to her ready wit in interpreting looks and gestures, and to her power to interest the starched, high-collared, precise, and generally verdant John. As I have always adhered to the plan we laid down when we first came to Paris, I abstained from taking advantage of these little helps to the English purchaser, especially as by so doing I obtained what I wanted at half the price which had to be paid where the article was served in our vernacular. However, on one occasion, I broke over this sensible regulation. I went out one morning by myself, leaving Franz employed at his easel. Happening to stroll through the aforesaid Passage, I observed, in one of the finest *boutiques*, the loveliest creature, t seemed to me, I ever beheld. Do not suppose this was alone sufficient to draw me in. It is an every-day occurrence, as you well know, this chancing on the "*most* beautiful." But, in the first place, the young girl was evidently fresh from the country. She knew nothing of her present occupation; she was not awkward, she could not be awkward, yet she did not seem at home in her new Parisian costume. She looked melancholy; in short, I was touched by her appearance. "Another victim!" I said to myself; how shocking to contemplate this poor innocent girl, so simple of heart, so modest, so beautiful, and think how soon she will be changed into a *Parisienne.*" I tried to throw off the idea: "it was but the old story; the country must supply the town; unfortunate, but necessary, and so forth: this young person appears melancholy, but it is only *la maladie du pays ;* she will soon be happy enough. Madame the manager treats her considerately; she is kind to her; a few days, and she will have her smiles again."

But days not a few passed, and no smile did I see. True, she was becoming acquainted with her business; she had learned to serve those who came with readiness; she seemed to have made rapid progress in learning what she had to do: but no smile, no "pleased alacrity," no quickening of the eye, no change of expression when the usual compliment was rendered by gay youth or handsome cavalier. The face was growing longer—perhaps more strictly beautiful; the cheek was losing its rose; the eyes appeared deeper, more subdued and thoughtful; indeed, the sight of her (I hardly know why, but I found myself passing the place daily) began to afflict me. Meanwhile, young men were crowding the *boutique;* for the singular beauty of the "charming grisette," her immobility, and the mystery which these created, became topics of conversation among the young Parisian "*lions*," as well as with a great many strangers. At this shop, I should have said, were exposed the words, "*English spoken;*" but the placard had only lately been posted, and I wondered who was the proficient in our unaccommodating tongue. So one morning, quite early, that I might have the fewer interruptions, I sauntered leisurely into the place, and inquired of the first one I saw, if she could speak English. "What you please to want, sir?" said the madame, coming up to me, and articulating with difficulty. I asked for some article not usually demanded, in order to test her knowledge. She hesitated, beckoned my heroine to her, and, leaving me in her charge, turned to serve a new-comer. I repeated my request in English, but did not attempt to explain by look or motion. The poor girl tried hard to divine what I would have; she bent forward, and I again repeated. It was interesting to observe how her natural intelligence strove to interpret what I was saying; the eyes grew full of meaning, and the countenance was roused from its repose; but it would not do. I had carefully avoided using any ordinary phrase; and as I stood still and spoke merely, it was no wonder that one who knew not a syllable of the language should fail to understand me. "*Pardon*," I said; "I thought some person here—(pointing to the placard)—spoke English." The girl turned with a distressed look to the madame, but she was busily engaged with her customer. Other grisettes there were, but my attendant made no appeal to them. "*Monsieur*," she finally said. "*je crains qu'on parle bien mal l'Anglais.*" This was uttered in a serious tone, and with an entire absence of pleasantry. Yet with what a graceful smile an ordinary French shop-girl would have said the same words, and have made you quite satisfied to remain and purchase whatever she chose to offer. I partly turned as if to depart, although I had no such intention, when the young girl placed her hand on a package of gloves that lay on the case, and looked at me inquiringly. I could perceive this was, on her part, an act of mere duty, lest the business of her employer should seem neglected. I said nothing, but allowed her to select me a pair. As I was engaged in fitting them, I cast a glance at her. The look she gave me in return was so sad, so heavy-hearted, so desolate, that I could not avoid saying to her in her own tongue, "*You seem very unhappy.*" A flush passed across her face, a tear forced its way into her eyes, and, before she could prevent it, dropped on her cheeks and rolled down her face. Her handkerchief was quickly applied, and she was calm and imperturbable as before. My tone was one not of gallantry, but of kindness, and it had taken her by surprise. Yet she said nothing—not a word; but she looked at me a moment intently, as if to question my motive in speaking to her;. but whether she was satisfied of it or not, I could not tell.

I did not seek to draw her into conversation, but took my leave as soon as I had paid for my purchase. I need not detail to you, my dear Partridge, how I finally succeeded in obtaining the confidence of Marie Laforet—for that was her name—and which put me in possession of her simple history. The young creature saw that I was painfully interested for her; besides, she knew not a soul in Paris to whom she could trust her sorrows. It seemed as if she would have died, could she not have spoken; and yet I fear you will be disappointed when I tell you there was nothing extraordinary in her story. No tale of a faithless lover or of cruel parents, of afflictions, or of harsh treatment by friends, or of any thing melo-dramatic. She informed me that she was a native of Burgundy, in the department of the Saone and Loire, and lived near the little town of *Charolles* with her mother, who owned a small farm of ten or fifteen acres, nearly all of which was vineyard.

The adjoining plot was occupied by Maurice Foligny and *his* mother, who were their nearest neighbours. Maurice was two-and-twenty, and Marie was his sweetheart. Their marriage had long been considered a settled affair, not only between the lovers, but by the old people themselves. In short, it was to take place at the coming vintage. During the spring, Marie's mother received a visit from an only sister who had gone to Paris in her youth, married a respectable shopkeeper, and succeeded, on his death, to his establishment. What had sent her so far away into the Departments to look up her sister whom she had not seen for twenty-five years, was difficult to imagine. Perhaps she felt a pride in displaying herself and her finery to her only surviving relative, and in acquainting her with the independent position she now held at the head of one of the handsomest shops in Paris; perhaps the motive might be attributed to that instinctive longing for one's kindred which steals over us after we have passed the boundary of middle life, gathering strength year by year, until with the aged it becomes engrossing, and at times almost unendurable. However this may be, Madame Duchamp —so she was designated—actually arrived and took up her quarters at the little farmhouse. Nothing was now heard of but Paris, Paris, Paris! No other place in tne universe could compare with it Every thing out of it was actually barbarous. Marie, to be sure, had a sweet face, was well-shaped, yet what a fright she was when disfigured by that *outré* dress! and when poor Maurice ventured into the presence of Madame, he was treated to such a frigid reception, that he never could be persuaded to come again; and Marie herself was overwhelmed by a shower of ridicule respecting the appearance of her lover. To shorten the tale, Madame Duchamp finally prevailed on her weak-minded sister, despite the entreaties and protestations both of Marie and Maurice, to send her daughter to Paris, that she might become a lady under the care and supervision of her experienced aunt. The troth of the young people was by no means broken; the shrewd Madame thought this to be quite unnecessary. She supposed Marie to be like most young girls, and depended on her forgetting her lover in a week after she should arrive in Paris, calculating the while on profiting largely by increased sales in consequence of having so beautiful a person in attendance. At the same time, her intentions were perhaps well meant; for she expected, without doubt, that her niece should succeed to her business, and inherit what she possessed. Meanwhile, poor Marie became utterly wretched; as I have described to you, she seemed slowly to wither away. She had been four months in Paris; she had not heard from Maurice, nor from her mother, except through *Madame*, and when she made these disclosures to me, was ready to sink into absolute despair. Poor, forlorn thing that she was! I went home revolving the matter in my mind. What was to be done? What could *I* do? I finally broke the subject to our friend Franz here: strange to say, up to this time I had kept the affair quite to myself. Now I wanted some one to consult with, and I knew Franz would appreciate the interest I took in the business. The result was, that we determined to make an incursion into Burgundy, work our way quite carelessly into the neighbourhood of Marie's home, and inspect the situation of things. You laugh, my dear boy, at this adventure— I know you do; you call it Quixotic. I cannot help it. I never commenced a journey with a more earnest purpose or a more cheerful heart; and if there was a sprinkling of romance in it, should it detract from the value of the object which we sought to compass? Obtaining from Marie such information as would enable us to find the desired locality without hinting the reason for the inquiry, my friend and I set off. It was not yet the season of the vintage, but the vine with its rich clusters already exhibited a luxuriant picture. We passed rapidly south, and at length reached *Charolles*. Here our reconnoisance commenced. We had no difficulty in finding the cottage of the Widow Laforet; and one afternoon, just at sunset, we entered her dwelling and asked for a draught of wine. I fancied there was an air of grief and of loneliness in her manner quite unnatural. She desired us to be seated, and provided for us the best her cottage afforded. Franz undertook to explain our movements. We were from Paris, he said, and were making a pleasure tour through this delightful part of France. At the mention of *Paris*, the Widow started, and her interest in what my friend was saying evidently increased.

"From Paris!" she exclaimed. "Then you must know my Marie!"

I could not help smiling at the poor woman's simplicity, but Franz preserved his gravity, and replied: "Perhaps—with whom does she live?"

"Ah!" responded the Widow Laforet, "you must have seen her; she is with Madame Duchamp; every body knows *Madame*."

"What!" demanded my friend, "Madame Duchamp, who keeps a shop in the *Passage des Panoramas?*"

"The very same, sir."

"And what did you say was the name of your daughter—Madame has several young girls with her?"

"Marie, sir: indeed, you could not mistake *my* Marie. You would know her among a thousand."

"She must mean Marie Laforet," said Franz, turning

to me with an air of indifference, as he proceeded to light his meerschaum.

"Ah, *mon Dieu!*" cried the poor widow; "it is, indeed, my own *petite* Marie. I was certain you knew her. Pray tell me all you can about her. She must be so happy in beautiful Paris, with every thing to delight her."

"I doubt if it is the same person," said Franz, stiffly.

"But I tell you that it is," said the other, with eagerness; "therefore go on; pray go on, sir."

"You will please describe your daughter," said my inexorable friend.

"To be sure. A fine shape, just my height; face round, fresh, with roses on her cheeks; fair skin; eyes —ah! so fine, so full, so gentle, so brown; hair, a chestnut; and her whole——"

"Not the same person," said Franz, again turning to me, and giving a puff of his meerschaum.

"But it is—I know that it is!" cried the Widow; there cannot be two Marie Laforets with my sister. Ah! I have forgotten: Marie is so much altered, so much improved in every way, that even her mother cannot describe her correctly. Just as my sister promised me—the dear, good one! But you will tell me how she looks now, just to please a foolish old woman—I know you will, sir."

"I doubt if it can be your daughter," answered Franz. "The Marie Laforet whom I have seen is, to be sure, about your height, and has chestnut hair and brown eyes· but her form seems to be wasted; her face is very pale and thin; her cheeks are colourless. Oh, no! it is not your little Marie;" and Franz drew some fresh tobacco from his pouch.

The Widow burst into tears. A vision of the true state of things passed over her.

It was now my turn. "I am sure," said I, "that the Marie whom we know *is* the daughter of our entertainer; the description agrees in every thing except in that wherein young people who are unhappy are most liable to change. It is true, that her cheeks are pale and hollow, and that she seems to be declining in health; otherwise it answers very well, depend upon it. My good woman," I continued, with severity, " you should see to your child."

"And you, too, know her!" said the Widow Laforet, not heeding my reproach, and looking up through her tears; "and you say she is miserable? Yes, miserable she must be—my own darling, precious Marie! Why did I trust her away from me? My sister should have told me of this. I suppose she hoped there would be a change for the better. Alas! I have not had a happy moment since she left me. Ah, what will poor Maurice say?"—and she continued her lamentations for several minutes.

"And who is Maurice?" inquired Franz.

"Maurice, sir, is a worthy lad, who is betrothed to my Marie. They were to be married the coming month; but this visit of my sister—alas! alas! it has ruined us all."

"And Maurice," said I; "how does he bear Marie's absence?"

"Indeed, sir, worse than any of us. Not a word has he heard from her, although he has sent her a great many letters; but he does not blame Marie, not he:— yet he does nothing but curse Madame Duchamp—God forgive him!—from one week's end to another. He now declares that as soon as the vintage is gathered he will go to Paris. Ah! the vintage this year will be so sad, when we were promising ourselves so much pleasure!"

"And why should you not have it?" said Franz abruptly, starting to his feet, and looking the Widow Laforet full in the face. "What is there to prevent you sending to Paris for Marie, and celebrating her nuptials with Maurice at the very time agreed upon?"

"But my sister," interposed the poor woman, timidly.

"*Diable!*" growled Franz; "would you sacrifice your own flesh and blood, body and soul, for fear of giving offence to——"

The sentence was cut short in an uncouth German guttural, which I should not care to have translated.

"But what shall I do?" continued the Widow: "how shall I manage it? I know nothing of the ways of the strange folks in Paris, and if I sent for Marie, my sister would not let her go, for she has been at large charges for her journey, and for dresses, and I know not for what else. Ah, I fear it cannot be; yet what will become of thee, *ma petite?*" And again she wept.

It was now evening, and we were urged to spend the night at the cottage. Franz shook his head, spoke of walking on to Charolles, but I overruled him, and he accepted the proffered hospitality. We were served with supper, and the good dame plucked for us, from her early fruitage, clusters of delicious grapes. I had sustained my part thus far tolerably well, but my heart was ready to burst at the sight of this poor woman attempting to be cheerful while she prepared our entertainment. As for our friend, I could not too much admire the admirable manner with which he had managed the interview. In the course of the evening I undertook to explain to the Widow Laforet the dangers of a life in Paris to a young girl situated like Marie, and was not long in convincing her that she had reason to rejoice that the atmosphere of the city agreed so ill with her child. Franz verified all I said by an abrupt, emphatic assent, so that before we retired, her only desire was to get her daughter away from such a place of abominations. Thus far our plan had succeeded admirably, and we went to sleep confident and sanguine. The next morning, the Widow asked our advice as to the best means of getting Marie back to her home. Her only embarrassment was how to brave her sister's displeasure, and how to make amends for the expenses she had incurred for her. These, to us, were minor considerations, for I knew the latter to be much exaggerated in the Widow's imagination, and as to the former, it seemed, under the circumstances, of no consequence whatever.

We at once proposed that Maurice should be sent for, and the dame accordingly went for him. As it was but a few steps, she soon returned, accompanied by Maurice Foligny, a fine, noble-looking fellow, of manly bearing, to whom, after being satisfied of his ready perception by a few minutes' conversation, I frankly stated our object in coming into the neighbourhood. When he fully understood it, he grasped the hands of each, and, without uttering a word, thus silently expressed his thanks.

I need not recount to you how my friend and I went back to Paris in high spirits, bearing a letter from the Widow Laforet to Marie, and also one to

Madame Duchamp, the latter being the joint production of Franz and myself, and written in a manner best adapted to effect our object without giving offence. Although mild and conciliatory, it was nevertheless decisive as to Marie's return, on the ground of her ill health and her mother's lonely situation, referring also to the promise of Madame Duchamp, which her sister at the last moment recollected to mention to me, that if, after a few months' trial, Marie or her mother were not content with the arrangement, the young girl should be sent back. I believe there was also a letter from Maurice to his betrothed, but as this is a point of little consequence, I will not speak positively.

The end of the whole business you may guess by this painting about which you were so inquisitive. *Madame* did not prove as obstinate as was expected. The fact is, she was pretty well convinced that Marie would never adapt herself to her new life, and consequently that the speculation was a failure; for, as the poor girl's health began to droop, even her mysterious demeanour ceased to attract attention. So she was sent home without more delay. The only astonishing part of the history is, how suddenly she recovered her health, her gayety, her plumpness, her colour, and the rich brown of her eyes, which had become so light and dull.

The next month came; we had pledged ourselves—Franz and I—to be present; and in the very heyday of the vintage, attended by a joyous company, Maurice and Marie were united in the little chapel which you see here; after which followed a dance upon the green, and a world of merrymaking. Our friend Franz seized the occasion to exhibit a happy proof of his art.

You were right, my dear Partridge: this *is* no fancy-sketch.

CHAPTER XIII.

PREFACE FOR CONCLUSION.

"'This will do, perhaps," said the literary friend to whom I submitted the foregoing pages; "but you have omitted one very important thing."

" What ?"

" The preface !"

" I hate prefaces."

" That may be; but I assure you the preface is as essential as the book."

" How so ?"

" It enables the reader to learn the scope and object of the work."

" Can these not be discovered in the perusal of it?"

" It is very doubtful. But this is not all: if one has a good preface, the critics can generally manage to get up an article without having to wade through the volume; and that, you know, is always a great recommendation for them. Indeed, I assure you, it is absolutely necessary for you to state in a preface the purpose you had in view in publishing your volume."

"What if I had none?"

"That is quite ridiculous. Every body nowadays writes with a design; *story* is the great medium for disseminating theories, philosophies, moralities——"

"And (*interrupting*) absurdities generally."

"You trifle. I was about to say, that whoever wishes to address the public in support of a favourite opinion, employs for a vehicle *dramatic fiction*. I have no doubt that very soon our clergymen will preach romances from the pulpit."

"Well!"

"You see, then, you must prepare a preface."

"But suppose I had no particular theory to bring forward in this book of mine?"

"Oh! you had, of course."

"But I say, suppose I had *not?*"

"Then, seriously, I advise you not to think of publishing it."

"Still, it contains descriptions of different phases of life?"

"Of no sort of consequence."

"It endeavours to pourtray the passions and emotions of the heart?"

"The object! the object!"

"It records actual reminiscences——"

"My friend, it is all very well, provided you have had what we call a 'persistent purpose' in what you have been doing."

"Is it not enough that I have written because I wanted to write? because it gave me pleasure? because it afforded me agreeable recreation after hours of professional labour?"

"I tell you it is *not* enough; to say that would be to say nothing. I warn you, for the last time, if you do not follow my advice, your book won't SELL!"

"What if it does not sell; have I not had the enjoyment of writing it?"

"Oh, indeed! if you are going to mount the high horse, I will bid you good-morning. I should like to hear what your publisher will say. Upon my word, here he comes!" [*Enter Publisher.*]

"My dear sir, I just stepped in to inform you that we are waiting for your preface."

"What did I tell you? You will believe *me* after this. I knew you would be forced to write one. But I will leave you together to settle the matter. Adieu!" [*Exit Literary Friend.*]

"I had decided not to have any preface."

"No preface?"

"To be sure, our friend who has just left advises it."

"He is a very judicious person. It would be safe, I am sure, to follow his suggestion."

"I suppose so; (*submissively;*) but how will this please you? I will write a preface, and insert it for the last chapter."

"That strikes me as rather a good idea. On the whole, I think it will *take*."

"Done!"

THE END.

By the Author of "Undercurrents."

ST. LEGER:

OR,

THE THREADS OF LIFE.

BY RICHARD B. KIMBALL.

New edition, 1 Vol. 12mo. Cloth. $1.25.

Extract from a private note to the author from

WASHINGTON IRVING.

"It is only within the last two or three days that I have taken your book in hand; and I now lay it down after having been deeply interested and delighted with the perusal. I do not pretend to criticise; it is not my forte: but I can *feel* when a work is good, and my feelings have been continually aroused and touched in the course of perusing your pages, while they are all calculated to set a man thinking. In a word, I find a power and beauty in your work and a fertility of invention (almost prodigal) which convince me we may confidently look for still better things at your hand.   *   *

"I shall be most happy to meet you and testify to you the great esteem and regard inspired by your writings."

Extracts from Notices of the Press.

"A Novel *sui-generis* in the annals of American literature."—*Phila. Journal.*

"Abounding in the most thrilling interest in narrative and in maxim."—*Metropolitan.*

"A book of great strength."—*N. Y. Evening Post.*

"A brilliant book, without a prototype in our literature."—*N. Y. Tribune.*

"A book of power."—*Boston Post.*

"A very extraordinary book."—*London Morning Post.*

"Here, there and everywhere, the author of 'St. Leger' gives exhibitions of passionate and romantic power."—*London Athenæum.*   **Turn over.**

THE BOOK OF THE DAY

Is the new Story by the Author of St. Leger,

UNDERCURRENTS

A

ROMANCE OF BUSINESS.

12mo., CLOTH, $1.25.

"One of the able and eloquent books of the day."—*Boston Post.*

"A genuine portraiture of the experience of financial embarrassmen in its relations to domestic life and personal character."—*Transcript.*

"For sharp-cut delineations of character hardly equalled by any contemporaneous writer."—*New York Tribune.*

'The book is an admirable success."—*Boston Daily Advertiser.*

"In all his portraitures the writer displays a power second only to bodily reproduction."—*Sunday Mercury.*

"Full of earnest words to business men."—*New York Evening Post.*

"Every line reads like a veritable history of an actual life." * * * "The characters are unmistakable historical pictures."—*Courier.*

NEW YORK: G. P. PUTNAM,
532 BROADWAY.   *Turn over.*

February, 1862.

LIST

OF

CHOICE BOOKS,

PUBLISHED BY

G. P. PUTNAM, 532 Broadway,

NEW YORK.

BOOKS ARE THE WINDOWS THROUGH WHICH THE SOUL LOOKS OUT. A HOUSE WITHOUT BOOKS IS LIKE A ROOM WITHOUT WINDOWS. NO MAN HAS A RIGHT TO BRING UP HIS CHILDREN WITHOUT SURROUNDING THEM WITH BOOKS, IF HE HAS THE MEANS TO BUY THEM. IT IS A WRONG TO HIS FAMILY. HE CHEATS THEM. CHILDREN LEARN TO READ BY BEING IN THE PRESENCE OF BOOKS. THE LOVE OF KNOWLEDGE COMES WITH READING AND GROWS UPON IT. AND THE LOVE OF KNOWLEDGE, IN A YOUNG MIND, IS ALMOST A WARRANT AGAINST THE INFERIOR EXCITEMENT OF PASSIONS AND VICES. * * * A LITTLE LIBRARY, GROWING LARGER EVERY YEAR, IS AN HONORABLE PART OF A YOUNG MAN'S HISTORY. IT IS A MAN'S DUTY TO HAVE BOOKS. A LIBRARY IS NOT A LUXURY, BUT ONE OF THE NECESSARIES OF LIFE.—*H. W. Beecher.*

NOW READY. PRICE FIVE CENTS, POST FREE.

Suggestions for Household Libraries

Of ESSENTIAL and STANDARD BOOKS, and the most economical mod obtaining them. With impartial lists of 250, 500, 1,000, and 1,250 volumes of the literature, and the most desirable editions. G. P. PUTNAM, Library Commi Agency, 532 Broadway, New York.

ALSO NOW READY, SENT POST FREE, FOR 35 CENTS.

G. P. Putnam's Classified General Catalogue

Of the MOST IMPORTANT WORKS in every department of Literature, En and American editions, with prices annexed. pp. 259.

Attractive and Standard Books,

PUBLISHED BY

G. P. PUTNAM, AGT.,

532 BROADWAY, New York.

The New Publications in Press or Just Ready.

THE

LIFE AND LETTERS

OF

WASHINGTON IRVING.

EDITED BY PIERRE M. IRVING.

PROBABLY FILLING THREE VOLUMES.

SUNNYSIDE EDITION, uniform with that edition of Irving's Works, large 12mo, cloth. Per vol., . . . $1.50
LARGE PAPER EDITION—(only 250 copies printed)—on laid paper, square 8vo, cloth. Per vol., ... 3.00

⁎ The first volume is now ready.

The Artist's

EDITION OF THE

Sketch Book,

WITH ABOUT 150 ORIGINAL DESIGNS.

A unique and superb volume, intended to surpass any similar publication yet produced in this country.

This volume has been more than two years in preparation, at large cost.

It is expected to be ready for the next season, and that most of the leading artists will be represented in it.

The paper and press-work, it is believed, have never been excelled in this country or in Europe.

Subscribers' names for choice and early copies received by the publishers, or by any bookseller.

The price of the volume, in rich morocco binding, will be $10.

WASHINGTON IRVING'S WRITINGS.

ALHAMBRA.

By WASHINGTON IRVING.

A Residence in the celebrated Moorish palace, the "Alhambra," with the historical and romantic legends connected therewith.

POPULAR EDITION, 1 vol., 12mo, green cloth, ...	$1.25
TINTED EDITION, Illustrated, cloth, gilt, extra, ...	2.00
" " " half calf, antique, ...	2.75
" " " half calf, extra, ...	2.75

"*The beautiful Spanish Sketch Book, the Alhambra.*"—W. H. PRESCOTT.

ASTORIA;

or, Anecdotes of an Enterprise beyond the Rocky Mountains. By WASHINGTON IRVING.

"*It is a book to put in your library, as an entertaining very well written account of savage life on a most extensive scale.*"—REV. SIDNEY SMITH.

This volume describes the first great explorations of the Oregon and adjacent territories, and the foundation of the great wealth of Mr. John Jacob Astor.

POPULAR EDITION, 12mo, green cloth, ...	1.50
TINTED EDITION, large 12mo, half calf, extra, . . .	2.75
" " " " half calf, antique, . . .	2.75

BONNEVILLE'S ADVENTURES.

The Adventures of Capt. Bonneville, U. S. A., in the Rocky Mountains and the Far West. From his own journals, and illustrated from other sources. By WASHINGTON IRVING.

POPULAR EDITION, 12mo, green cloth, ...	1.25
TINTED EDITION, large 12mo, half calf, extra, . . .	2.75
" " " " half calf, antique, . . .	2.75

"*Full of exciting incident,*   *   * *with the power and the charms of romance.*"—CHANCELLOR KENT.

BRACEBRIDGE HALL,

or the Humourists. By WASHINGTON IRVING.

POPULAR EDITION, 12mo, green cloth,	$1.25
TINTED EDITION, Illustrated, large 12mo, gilt, extra,	2.00
" " " half calf, extra,	2.75
" " " half calf, antique,	2.75
ILLUSTRATED (large paper) EDITION, 8vo, cloth,	3.50
" " " " 8vo, gilt edges,	4.00
" " " " 8vo, morocco, extra,	6.50
" " " " 8vo, morocco, antique,	6.50

"*It is very ill-natured, however, to object to what has given us so much pleasure.*"—LORD JEFFREY.

CHRONICLES of the CONQUEST of GRANADA.

By WASHINGTON IRVING.

POPULAR EDITION, 12mo, green cloth,	1.25
TINTED EDITION, large 12mo, half calf, antique,	2.75
" " " half calf, antique,	2.75

"*It has superseded all further necessity for poetry, and, unfortunately for me, for history.*"—W. H. PRESCOTT.

COLUMBUS AND HIS COMPANIONS.

The Life and Voyages of Christopher Columbus: to which are added those of his Companions. By WASHINGTON IRVING.

POPULAR EDITION, 3 vols., 12mo, green cloth,	4.00
TINTED EDITION, Illustrated, 3 vols., large 12mo, half calf ant.,	8.00
" " " " " half calf, extra,	8.00
OCTAVO EDITION, Illustrated, 3 vols., 8vo, cloth, extra,	10.00
" " " " 8vo, half calf, extra,	15.00
" " " " 8vo, half calf, antique,	15.00
" " " " 8vo, full calf, extra,	18.00

"*In treating this happy and splendid subject, Mr. Irving has brought out the full force of his genius.*"—ALEX. H. EVERETT.

"*The noblest monument to the memory of Columbus.*"—W. H. PRESCOTT.

"*It will supersede all other works on the subject, and never be itself superseded.*"—LORD JEFFREY.

CRAYON MISCELLANY.

Comprising a Tour on the Western Prairies; Abbottsford and Sir Walter Scott, and Newstead Abbey and Lord Byron. By WASHINGTON IRVING.

POPULAR EDITION, 1 vol., 12mo,	1 25
TINTED EDITION, large 12mo, half calf, antique,	2.75
" " " half calf, extra,	2.75

Diedrich Knickerbocker's

HISTORY OF NEW YORK, from the beginning of the World to the end of the Dutch Dynasty: Containing, among many surprising and curious matters, the Unutterable Ponderings of Walter the Doubter; the Disastrous Projects of William the Testy; and the Chivalric Achievements of Peter the Headstrong—the Three Dutch Governors of New Amsterdam; being the only Authentic History of the Times that ever hath been or ever will be published. [By WASHINGTON IRVING.]

POPULAR EDITION, 1 vol., 12mo, ...	$1.25
TINTED EDITION, with Plates and Cuts, large 12mo, gilt, extra,	2.00
" " " " " half calf, extra,	2.75
" " " " " half calf, ant.,	2.75
ILLUSTRATED (large paper) EDITION, 8vo, cloth, . . .	3.50
" " " " " mor., extra, . .	6.50
" " " " " mor., antique, . .	6.50
" " " " " calf, antique, red edges,	6.50

"*The most excellently jocose History of New York. * * * Our sides have been absolutely sore with laughing.*"—SIR WALTER SCOTT.

"*A book of unwearying pleasantry.*"—EDWARD EVERETT.

"*The most elaborate piece of humor in our literature.*"—H. T. TUCKERMAN.

"*Manly, bold, and so altogether original, without being extravagant, as to stand alone among the labors of men.*"—BLACKWOOD'S MAGAZINE.

Mahomet and his Successors.

By WASHINGTON IRVING.

POPULAR EDITION, 2 vols., 12mo, green cloth, ...	2.50
TINTED EDITION, EXTRA PLATES, 2 vols., half calf, extra, .	5.50
" " " " " half calf, antique, .	5.50

Oliver Goldsmith: a Biography.

By WASHINGTON IRVING.

POPULAR EDITION, 1 vol., 12mo, green cloth, ...	1.25
TINTED EDITION, Illustrated, large 12mo, gilt, extra, . .	2.00
" " " " half calf, antique, .	2.75
" " " " half calf, extra, .	2.75

"*I have read no biographical work which carries forward the reader so delightfully. * * I know of nothing like it.*"—WM. CULLEN BRYANT.

SALMAGUNDI.

By WILLIAM IRVING, JAMES K. PAULDING, and WASHINGTON IRVING.

POPULAR EDITION, 1 vol., 12mo, green cloth,	$1.25
TINTED EDITION, 1 vol., large 12mo, green cloth,	1.50
" " " " half calf, extra,	2.75
" " " " half calf, antique,	2.75
LARGE PAPER EDITION, 8vo, morocco, extra,	6.00
" " " " morocco, antique,	6.00

"*Full of entertainment, with an infinite variety of characters and circumstances, and with that amiable, good-natured wit and pathos,*" &c.—R. H. DANA.

SKETCH BOOK
OF GEOFFREY CRAYON, GENT.

By WASHINGTON IRVING.

POPULAR EDITION, 1 vol., 12mo, green cloth,	1.25
TINTED EDITION, with Woodcuts, cloth gilt, extra,	2.00
" " " half calf, antique,	2.75
" " " half calf, extra,	2.75

[☞ SEE *the Artist's Edition of the Sketch Book.*]

"*It is positively beautiful.*"—SIR WALTER SCOTT.
"*This exquisite miscellany.*"—J. G. LOCKHART.

TALES OF A TRAVELLER.

By WASHINGTON IRVING.

POPULAR EDITION, 1 vol., 12mo, green cloth,	1.25
TINTED EDITION, large 12mo, with cuts, gilt, extra,	2.25
" " " " half calf, antique,	2.75
" " " " half calf, extra,	2.75
ILLUSTRATED (large paper) EDITION, 8vo, cloth, extra,	3.50
" " " " 8vo, morocco, extra,	6.50
" " " " 8vo, morocco, antique,	6.50

WOLFERT'S ROOST.

By WASHINGTON IRVING.

POPULAR EDITION, 1 vol., 12mo, green cloth,	1.25
TINTED EDITION, large 12mo, gilt, extra,	2.25
" " " " half calf, extra,	2.75
" " " " half calf, antique,	2.75

"*We envy those who read these tales and sketches of character for the first time.*"—WESTMINSTER REVIEW.

WASHINGTON.

THE LIFE OF GEORGE WASHINGTON.

By WASHINGTON IRVING.

"His Life of Washington is a marvel."—GEO. BANCROFT.

I. POPULAR EDITION, 5 vols., 12mo, green cloth, . . . $7.00
  "   "   "   12mo, sheep, extra, . . 8.50

II. SUNNYSIDE EDITION, with plates, 5 vols., 12mo, cloth, . . 8.00
  "   "   "   " 12mo, half calf, extra, 12.50
  "   "   "   " 12mo, half calf, ant., 12.50
  "   "   "   " 12mo, hf. mor., gt. ed., 14.00
  "   "   "   " 12mo, full calf, extra, 15.00
  "   "   "   " 12mo, full mor. gt. ed., 17.00

III. UNION EDITION, steel plates, 5 vols., small 8vo, cloth, uncut, 8.50
  "   "   "   "   "   " hf. cf., extra, 14.00
  "   "   "   "   "   " hf. cf., ant., 14.00
  "   "   "   "   "   " full cf., ext., 16.00
  "   "   "   "   "   " full mor. ex., 18.00

IV. LIBRARY EDITION, octavo, 5 vols., cloth, ... 10.00
  "   "   "   " sheep, . . . 12.50
  "   "   "   " half calf, antique, . . 16.00
  "   "   "   " half calf, extra, . . 16.00

V. MOUNT VERNON EDITION, 100 Steel Plates and 40 Wood Cuts, on tinted paper. A new and superb edition. 5 vols., square 8vo, cloth, ... 17.00
  half calf, extra, . . 24.00
  half calf, antique, . . 24.00
  half morocco, gilt tops, 24.00
  full calf, extra, . . 28.00
  full morocco, gilt edges, 30.00
  full morocco, gilt, extra, 30.00

VI. ILLUSTRATED (large paper) EDITION, 5 vols., royal 8vo, half calf, antique, ... 30.00
  half mor., extra, 30.00
  full mor., extra, 36.00

⁎ *In this edition the pages are enclosed in lines.*

VII. AMATEURS' (Quarto) EDITION, with 102 Steel Plates, proofs on India Paper, 5 vols., quarto, morocco, extra, . . 100.00

⁎ *Of this superb edition only 110 were printed, and but 3 copies remain unsold.*

AN EDITION FOR CANVASSERS is also printed in ONE large volume, double columns, with Twenty Steel Plates. Price, $5, in cloth. Also, in Twenty-Six Numbers, with Fifty-Two Plates. Price, 25 cents per number—making two handsome volumes. Price, ... 7.50

WASHINGTON IRVING'S WHOLE WORKS,

Including LIFE OF WASHINGTON, SALMAGUNDI, and all the Miscellaneous Writings.

Complete in Twenty-two Volumes, 12mo.

SUNNYSIDE EDITION, Tinted Paper, with Vignettes,

(a)	22 vols., 12mo, extra cloth, ...	$30.00
(b)	" " extra sheep, ...	35.00
(c)	" " half calf, plain, ...	47.00
(d)	" " half calf, extra, ...	50.00
(e)	" " half calf, antique, ...	50.00
(f)	" " half morocco, gilt edges, . . .	55.00
(g)	" " full calf, extra, ...	60.00
(h)	" " full calf, antique, ...	60.00
(i)	" " full morocco, extra, ...	70.00

WASHINGTON IRVING'S WRITINGS, exclusive of "WASHINGTON" and including "SALMAGUNDI." *Complete in 17 Volumes.*

(k)	SUNNYSIDE EDITION, cloth, extra, ...	22.00
(l)	" " half calf, plain, ...	35.00
(m)	" " half calf, antique, ...	38.00
(n)	" " half calf, extra, ...	38.00
(o)	" " full calf, extra, ...	44.00

"*I cannot hesitate to predict for him a deathless renown.* * * *He whose works were the delight of our fathers and are still ours, will be read with the same pleasure by those who come after us.*"—WM. CULLEN BRYANT.

WASHINGTON IRVING'S WRITINGS, NATIONAL EDITION. Illustrated. On superfine tinted paper.

☞ *This edition is printed only for Subscribers.*

Complete in Twenty-two vols., extra black cloth, bevelled edges.

✱✱ These may be subscribed for to be delivered monthly or oftener, for one dollar and a half per month.

(p)	A Complete Set thus, costs	33.00
(q)	The Same Edition, cloth, uncut, ...	33.00
(r)	" " extra green cloth, cut edges, not bevelled,	33.00
(s)	" " half calf, extra, ...	55.00
(t)	" " half calf, antique, ...	55.00

WASHINGTON IRVING'S WRITINGS, LARGE PAPER EDITION. Beautifully Printed, and mostly Illustrated. [*Only 100 sets printed.*] Twenty-two volumes,

8vo., cloth, extra, ...	70.00
The same, in half calf or half morocco, ...	110.00

✱✱ Early application should be made for these, as only 100 sets are printed, and no more will be done in this style. Thirteen volumes are now ready. Remainder in *September.*

Washington Illustrations.

Comprising 102 Steel Plates, Proof-impressions on India Paper, including 70 Portraits of eminent men of the Revolution, and 30 Historical Scenes by Darley, Trumbull, &c.

Quarto, in portfolio, ... $25.00
Any Plate may be had separately, price 40
WASHINGTON ILLUSTRATIONS to match the 8vo edition, 1 vol., cloth, 6.00

Bayard Taylor's Writings.

" There is no romance to us quite equal to one of Bayard Taylor's books of travel."—HARTFORD REPUBLICAN.

Eldorado;
OR, ADVENTURES IN THE PATH OF EMPIRE, (Mexico and California.)

12mo, cloth, ... $1.25

Central Africa;
LIFE AND LANDSCAPE FROM CAIRO TO THE WHITE NILE.

Two plates and cuts, 12mo, ... 1.25

Greece and Russia;
WITH AN EXCURSION TO CRETE.

Two plates, 12mo, ... 1.25

Home and Abroad;
A SKETCH BOOK OF LIFE, SCENERY, AND MEN.

Two plates, 12mo, cloth, ... 1.25

Home and Aboad—(Second Series.)
A new volume, just published, (1862.)

Two plates, 12mo, cloth, ... 1.25

India, China, and Japan.
Two plates, 12mo, cloth, ... 1.25

BAYARD TAYLOR'S WRITINGS—*Continued.*

LANDS OF THE SARACEN;
Or, PICTURES OF PALESTINE, ASIA MINOR, SICILY, AND SPAIN. With two plates, 12mo, cloth, ... $1.25

NORTHERN TRAVEL;
SUMMER AND WINTER PICTURES OF SWEDEN, DENMARK, AND LAPLAND.
Two plates, 12mo, cloth, ... 1.25

VIEWS A-FOOT;
OR, EUROPE SEEN WITH KNAPSACK AND STAFF.
12mo, ... 1.25

ROMANCE OF AMERICAN LIFE.
(In Press.)

BAYARD TAYLOR'S TRAVELS.
POPULAR EDITION, complete in 9 vols., 12mo, cloth, extra, . 12.00
"   "   "   "   "   sheep, extra, . 14.00
"   "   "   "   "   half calf, extra, . 20.00
"   "   "   "   "   half calf, ant., . 20.00

BAYARD TAYLOR'S PROSE WRITINGS.
CAXTON EDITION, printed for subscribers. In 10 vols., on tinted paper. Issued monthly. Price, per vol., . . 1.50

The volumes are issued as follows:

I. HOME AND ABROAD.
II. VIEWS A-FOOT.
III. HOME AND ABROAD. SECOND SERIES.
IV. ELDORADO.
V. CENTRAL AFRICA.
VI. LANDS OF THE SARACEN.
VII. INDIA, CHINA, AND JAPAN.
VIII. NORTHERN TRAVEL.
IX. GREECE AND RUSSIA.
X. ROMANCE OF AMERICAN LIFE. (New.)

A Magnificent and Valuable Work.

THE ARCHITECTURAL INSTRUCTOR;

Containing a History of Architecture from the Earliest Ages to the Present Time; illustrated with nearly 250 Engravings of Ancient, Mediæval, and Modern Cities, Temples, Palaces, Cathedrals, and Monuments; also, the Greek and Early Roman Classic Orders, their principles and beauties; with a large number of Original Designs of Cottages, Villas, and Mansions of different sizes, accompanied with practical observations on Construction, with all the important details, on a scale sufficiently large and definite to enable the Builder to execute with accuracy; and further Designs of Churches, Monuments, and Public Buildings; together with a Glossary of Architectural Terms; the whole being the result of more than Thirty Years' Professional Experience. By MINARD LAFEVER, Architect.

1 vol., large quarto, half morocco, gilt tops, ... $16.00

** *This handsome volume is, probably, the most comprehensive single volume on architecture ever published—embracing a full History of Architecture from the earliest times, and also a complete treatise on its theory and practice, with designs for modern houses, public buildings, churches, &c. It should be in every good library for general reference, as well as in the hands of every architect. Mr. Lafever was one of the most eminent and thorough architects in the country.*

AMERICAN HISTORICAL AND LITERARY CURIOSITIES;

Consisting of Facsimiles of Original Documents relating to the Events of the Revolution, &c., &c., with a variety of Reliques, Antiquities, and Modern Autographs. Collected and Edited by JOHN JAY SMITH and JOHN F. WATSON.

Sixth edition, with improvements and additions.

Large quarto, cloth, gilt tops, ... 8.00
" " half morocco, gilt edges, ... 10.00

A SECOND SERIES (COMPLETE IN ITSELF) OF

American Historical and Literary Curiosities;

Consisting of Facsimiles relating to Columbus, and Original Documents of the Revolution, with Reliques, Autographs, &c. Edited by JOHN JAY SMITH.

Quarto, cloth, ... 8.00
" half morocco, gilt edges, ... 10.00

WASHINGTON PORTRAITS.

The Character and Portraits of Washington. By HENRY T. TUCKERMAN. With 12 Portraits, proofs on India Paper. Only 150 printed.

Quarto, cloth, ... 6.00
" in a portfolio, ... 6.00

Mr. Lossing's Edition of Trumbull's McFingal.

(AN EPIC POEM OF THE REVOLUTION.) With copious Notes. Large paper. Only 100 copies printed.

Royal 8vo, sewed, ... 4.00

New Romance of Real Life.

UNDERCURRENTS OF WALL STREET;

A ROMANCE OF BUSINESS. By RICHARD B. KIMBALL, Author of "ST. LEGER." 1 vol., 12mo, ... $1.25

ST. LEGER; THE THREADS OF LIFE.

By RICHARD B. KIMBALL. Sixth Edition, 12mo, cloth, . 1.25

"*I find great power and beauty in your book, and a fertility of invention almost prodigal.*"—WASHINGTON IRVING.

"*A brilliant book, full of suggestions of wisdom.*"—NEW YORK TRIBUNE.

"*A book of great strength.*"—N. Y. EVENING POST.

"*Who is the author of this powerfully-written book?*"—PHILADELPHIA EVENING BULLETIN.

"*Here, there, and everywhere, the author gives exhibitions of passionate and romantic power.*"—LONDON ATHENÆUM.

"*A very extraordinary book.*"—LONDON MORNING POST.

The American Robinson Crusoe.

KALOOLAH;

A ROMANCE. By W. STARBUCK MAYO, M. D.

In 1 vol., 12mo, 512 pages, with illustrations by Darley, . . 1.50

"*The most singular and captivating romance since Robinson Crusoe.*"—HOME JOURNAL.

"*By far the most fascinating and entertaining book we have ever read since we were fascinated by the graceful inventions of the Arabian Nights.*"—DEMOCRATIC REVIEW.

*** Mr. Washington Irving considered Kaloolah to be a work of decided genius, and of extraordinary interest. The description of the magnificent city of Killoam, the capital of the great nation in the heart of Africa, includes a capital commentary and satire on the municipal regulations of American cities, and especially of New York.

ANGLO-SAXON GRAMMAR.

A GRAMMAR OF THE ANGLO-SAXON LANGUAGE. By L. F. KLIPSTEIN. 12mo, cloth, ... 1.25

*** *The best Anglo-Saxon Grammar in the language.*

MANUAL OF POLITICAL ECONOMY.

By E. PESHINE SMITH. 12mo, cloth, ... 1.25

*** *Used as a text-book at Princeton and other colleges, and well adapted for popular reading.*

ELEMENTS OF GEOLOGY.

Intended for the use of Students. By SAMUEL ST. JOHN, Professor of Chemistry and Geology in the College of Physicians and Surgeons, New York. Eleventh Edition. 12mo, cloth, . 1.00

The World's Progress;

A DICTIONARY OF DATES: *From the Creation to A.D. 1861.* Edited by G. P. Putnam, A. M., Hon. Mem. Conn. Historical Society, Wisconsin Historical Society, &c.

[An entirely new edition, with copious additions; including the most important facts in the History of the World, down to the Inauguration of Abraham Lincoln.]

The volume now contains more than ONE MILLION FACTS, on all topics connected with the progress of society, from the earliest period to the present, arranged for convenient reference.

1 vol., large 12mo, over 800 pages,	$2.00
half calf, extra,	3.00
A few copies on large paper, interleaved,	4.00
half morocco,	6.00

"*A more convenient literary labor-saving machine than this excellent compilation, can scarcely be found in any language.*"—NEW YORK TRIBUNE.

"*It has been planned so as to facilitate access to the largest amount of useful information in the smallest possible compass.*"—BUFFALO COURIER.

"*The best manual of the kind that has yet appeared in the English language.*"—BOSTON COURIER.

"*An exceedingly valuable book; wellnigh indispensable to a very large portion of the community.*"—COURIER AND ENQUIRER.

"*It is absolutely* ESSENTIAL *to the desk of every* MERCHANT *and the table of every* STUDENT *and* PROFESSIONAL MAN."—CHRISTIAN ENQUIRER.

Uniform with the Above.

Cyclopedia of Universal Biography.

By PARKE GODWIN, Esq., Author of the "History of France." New edition, with continuation to 1861.

1 vol., large 12mo,	2.00
The same in 8vo,	4.00

The Best and Most Complete Edition.

Papers for the People.

A Series of Popular and Instructive Papers on History, Archæology, Biography, Science, Industrial and Fine Arts, Civilization, Fiction, Personal Narrative, and other branches of Elegant Literature. Edited by ROBERT CHAMBERS.

12 vols., 12mo. Bound in 6. Red cloth, in a box. Price,	6.00

*** These admirable volumes comprise about 8,500 large and substantial pages, including a great variety of valuable and entertaining information on the subjects above indicated, interspersed with attractive narratives and tales, and all written with eminent ability by some of the most competent authors of the day.

The work is, in fact, a little LIBRARY IN ITSELF, eminently worthy of a place in every household, and attractive alike to young and old.

It is probably the CHEAPEST WORK (considering its eminent excellence and value) ever published.

Every FAMILY, and every PUBLIC LIBRARY and SCHOOL LIBRARY, should have a copy.

The above work is also issued in parts, royal 8vo, each part containing two distinct works, with a steel plate,	.25

In Quarto Numbers, elegantly printed, each containing Two Fine Portraits, with letterpress. Published Semi-Monthly, or oftener. Price, 25 Cents each No.

Heroes and Martyrs;
NOTABLE MEN OF THE TIME.

BIOGRAPHICAL SKETCHES
OF THE
MILITARY & NAVAL LEADERS, STATESMEN, & ORATORS,
DISTINGUISHED IN THE AMERICAN CRISIS OF 1861.

Edited by FRANK MOORE,

With Portraits on Steel, from Original Sources.

Heroes and Martyrs Illustrated.

"A work of very attractive interest and value is in preparation by Mr. G. P. PUTNAM, the publisher of the 'Rebellion Record.' In a series of numbers handsomely printed in quarto, he proposes to give personal sketches of the 'Notable Men of the Time, and the Heroes and Martyrs of the War,' illustrated with fine portraits on steel, from original photographs. Not only the generals and military leaders, but the YOUNG MEN of genius and promise—such as Greble, Ellsworth, Winthrop, Lieut. Putnam, and others, distinguished by character and talents in this great struggle—and the leading statesmen and orators of all parties, will be fully and fairly represented by faithful and accurate biographies. The work is to be a serial, but will form, when completed, a handsome volume of permanent interest."—*N. Y. Commercial Advertiser.*

HEROES, MARTYRS, & NOTABLE MEN, 1861-2.

Among the Portraits already engraved, or in progress, are the following:

[Others will be included as they become distinguished.]

ABRAHAM LINCOLN,
WM. H. SEWARD,
SALMON P. CHASE,
SIMON CAMERON,
GIDEON WELLES,
CHARLES SUMNER,
HENRY WILSON,
DANIEL S. DICKINSON,
EDWARD EVERETT,
JOHN P. KENNEDY,
JOHN J. CRITTENDEN,
ANDREW JOHNSON,
JOSEPH HOLT,
PARSON BROWNLOW
GEO. D. PRENTICE,
GOV. WM. SPRAGUE,
GOV. E. D. MORGAN,
WM. CULLEN BRYANT,
JOHN LOTHROP MOTLEY,
WENDELL PHILLIPS,

Rev. Dr. BELLOWS,
Rev. Dr. TYNG,
FRED. LAW OLMSTEAD.

Capt. WARD,
Commodore STRINGHAM,
Commodore DUPONT,
Capt. CHARLES WILKES.

Lt.-Gen. WINFIELD SCOTT,
Maj.-Gen. G. B. McCLELLAN
Maj.-Gen. JOHN E. WOOL,
Maj.-Gen. H. W. HALLECK,
Maj.-Gen. J. C. FREMONT,
Maj.-Gen. N. P. BANKS,
Maj.-Gen. B. F. BUTLER,
Maj.-Gen. JOHN A. DIX,
Brig.-Gen. R. ANDERSON,
Brig.-Gen. NATH. LYON,
Brig.-Gen. W. J. ROSECRANS,
Brig.-Gen. A. E. BURNSIDE,

Brig.-Gen. W. T. SHERMAN,
Brig.-Gen. E. D. BAKER,
Brig.-Gen. LOUIS BLENKER,
Brig.-Gen. MANSFIELD,
Brig.-Gen. McDOWELL,
Brig.-Gen. F. W. LANDER,
Brig.-Gen. F. SIEGEL,
Brig.-Gen. T. F. MEAGHER,
Col. E. E. ELLSWORTH,
Col. CORCORAN,
Maj. THEO. WINTHROP,
Lieut. SLEMMER,
Lieut. GREBLE,
Lieut. PUTNAM.

JEFFERSON DAVIS,
A. H. STEPHENS,
Rt. Rev. Bp. POLK,
P. G. T. BEAUREGARD,
ROBERT S. GARNETT.

THE LIBRARY

"THE PRESIDENT OF THE UNITED STATES says this work "*will soon be indispensable.*"

REBELLION RECORD,
A DIARY OF AMERICAN EVENTS.

With Illustrative Documents and Narratives, Rumors, Incidents, Poetry, Anecdotes, &c.,

IN SEPARATE DIVISIONS.

Edited by **FRANK MOORE**, Author of "Diary of American Revolution."

With an Introductory View of the Great Issues before the Country, and the Causes of the Rebellion,

By **EDWARD EVERETT.**

Also, important articles by MOTLEY, SEWARD, KENNEDY, HOLT, &c., &c.
SUMNER, DICKINSON, STEPHENS, DAVIS, &c., &c.,
Prepared (with additions) for this work. With a copious INDEX, etc., etc.

VOL. I. CONTAINS

A COLORED MAP OF THE UNITED STATES, and PORTRAITS ON STEEL of

Gen. SCOTT, Gen. BUTLER, Gen. CAMERON,
Gen. MCCLELLAN, Gen. ANDERSON, Gen. LYON,
Gen. FREMONT, Gen. DIX, Gov. SPRAGUE,
President LINCOLN, and JEFFERSON DAVIS.

The Second Vol. contains Fine Steel Portraits of

Com. STRINGHAM, Maj.-Gen. WOOL, Maj.-Gen. BANKS,
Brig.-Gen. BLENKER, Brig.-Gen. MCCALL, Brig.-Gen. MCDOWELL,
Brig.-Gen. MANSFIELD, Brig.-Gen. ROSECRANS, Brig.-Gen. LANDER,
Rt. Rev. Bp. POLK, Gen. BEAUREGARD, Brig.-Gen. BURNSIDE.

PRICE OF EACH VOLUME:

Cloth, $3.75; *Sheep*, $4; *Half Calf, ant.*, $5; *Half Mor.*, $5.

This work will be published as heretofore—weekly and monthly. Weekly Nos. at 10 cents; Monthly Parts, Illustrated, 50 cents. The Illustrations for the Weekly Nos. will be published in Two Nos. at 30 cents each, making the price for weekly and monthly editions the same—viz., $3.00 for each Vol. each six months. Covers for binding each, in cloth, 25 cents.

" * * * I consider the 'Record' a very valuable publication. * * "—*Edward Everett.*

"Every one who wishes a complete record of the stirring events now transpiring, should procure this weekly serial. In a small compass it gives the contents of a dozen daily newspapers."—*N. Y. Independent.*

"It is quite indispensable for reference, and forms one of the most remarkable specimens of *current history* ever published. We advise those who would preserve and ponder the authentic chronicle of 'the Second War of American Independence,' to possess themselves of this valuable and interesting serial."—*Boston Transcript.*

"We cannot speak too highly of the industry and sound judgment the work displays."—*N. Y. Daily Times.*

THE PULPIT RECORD;

A COLLECTION OF SERMONS, by Eminent Divines, North and South, with reference to the American Crisis of 1861–'62.

Parts I. and II., 8vo, each 50
The same, in 1 vol., 8vo, cloth, ... 1.50

Pure and Entertaining Reading for the Household.

THE

Aldine Edition of Thomas Hood.

SONG OF A SHIRT. [BEING THE FIRST EVER PUBLISHED COMPLETE.] BRIDGE OF SIGHS.

(Following the NATIONAL EDITION OF IRVING.)

MISS KILMANSEGGE. DREAM OF EUGENE ARAM.

The Works of Thomas Hood,

WHIMS AND } IN PROSE AND VERSE, { ODDITIES.

EDITED BY EPES SARGENT.

ILLUSTRATED WITH WOODCUTS FROM HOOD'S OWN DESIGNS, AND WITH VIGNETTES ON STEEL.

Elegantly Printed on Superfine Tinted Paper, in small octavo.
THE WORKS ARE ISSUED IN MONTHLY VOLUMES.

Each, in Cloth, extra, $1.50; Uncut, $1.50; Half Calf, extra, $2.50; Half Calf, antique, $2.50.

"Subtle fancy, lively wit, copious language, and mellow versification, are the undoubted qualities of Hood as a poet. But, besides, there are two or three moral peculiarities about him as delightful as his intellectual; and they are visible in his serious as well as lighter productions. One is his constant lightsomeness of spirit and tone. * * * But best of all in Hood is that warm humanity which beats in all his writings. His is no ostentatious or systematic philanthropy; it is a mild, cheerful, irrepressible feeling, as innocent and tender as the embrace of a child."—*Tait's Magazine.*

"Hood's verse, whether serious or comic—whether serene, like a cloudless autumn evening, or sparkling with puns, like a frosty January midnight with stars, was ever pregnant with materials for thought."—*D. M. Moir.*

"In the whole range of his works there is not a single line of immoral tendency, or calculated to pain an individual."—*Literary Gazette.*

"The master-spirit of modern whim and drollery."—*London Athenæum.*

His writings will "make the thoughtful wiser, and the unthinking merrier."—*New Monthly Magazine.*

** This Edition is sold only to Subscribers.

To follow the NATIONAL EDITION OF IRVING.

The Caxton Edition of BAYARD TAYLOR'S WORKS.

Now Ready, simultaneously with the Aldine Edition of Thomas Hood, the first volume of the new and beautiful edition of

THE

PROSE WRITINGS

OF

BAYARD TAYLOR.

In 10 vols., small 8vo, elegantly printed on superfine tinted paper,

With Several STEEL VIGNETTES. Price $1 50 per vol., in extra cloth.

"Views A-Foot,"

The first volume of Bayard Taylor's Travels, was issued in 1847, since which nearly 40,000 copies have been called for, and the demand for this and the later volumes continues to be remarkably large and constant. Probably no similar publications by a young author were ever received with such general favor, or reached so large a sale. Excepting only Kane's Arctic Voyages, perhaps no other American book of Adventure has been more popular with the people at large.

In accordance with suggestions from various quarters, and to meet the improved standard of taste and excellence in book-making, the publisher will issue these popular volumes at monthly intervals, printed on a superior tinted paper, with vignette engravings; and tastefully bound in vols., ranging with Irving and Hood. The volumes will be issued in the following order:

HOME AND ABROAD.	CENTRAL AFRICA.
HOME AND ABROAD,	LANDS OF THE SARACEN.
(Second Series—A new vol.)	INDIA, CHINA, AND JAPAN.
VIEWS A-FOOT.	NORTHERN EUROPE.
ELDORADO.	GREECE AND RUSSIA.

A ROMANCE OF AMERICAN LIFE (New.)

New York: G. P. PUTNAM, 532 Broadway.

Subscriptions received by the present agents for the Subscription edition of Irving.

www.ingramcontent.com/pod-product-compliance
Lightning Source LLC
Chambersburg PA
CBHW031934230426
43672CB00010B/1922